ENVIRONMENTAL MODELLING: NEW RESEARCH

Environmental Modelling: New Research

Paul N. Findley
Editor

Nova Science Publishers, Inc.
New York

For permission to use material from this book please contact us:
Telephone 631-231-7269; Fax 631-231-8175
Web Site: http://www.novapublishers.com

LIBRARY OF CONGRESS CATALOGING-IN-PUBLICATION DATA

Environmental modelling : new research / Paul N. Findley, editor.
p. cm.
ISBN 978-1-60692-034-3 (hardcover)
1. Environmental sciences--Mathematical models. I. Findley, Paul N.
GE45.M37E5935 2009
577.01'5118--dc22

2008037503

Published by Nova Science Publishers, Inc. ✦ New York

CONTENTS

PREFACE

Environment models seek to re-create what occurs during some event in nature. It is much easier and practical to create computer models to run certain experiments than it is to go out and do the same experiment again and again. Computer models take equations which were usually formulated through testing under natural conditions, and put them into computer programs where they can be run quickly and easily. A model can then output the results of doing these equations into a form which can be output to a screen for the user to view. The aim is to improve the capacity to represent, understand, predict or manage the behaviour of environmental systems at all practical scales. This new book presents the latest research from around the globe.

As presented in the Expert Commentary, in recent decades, digitally georeferenced data and Geographic Information Systems (GIS) have played a growing role in human exposure assessment for environmental epidemiologic studies. Despite the increased use of GIS in exposure assessment research, spatio-temporally varying datasets, such as daily activity spaces, residential histories, and time-varying maps of environmental contaminants are poorly characterized in the GIS environment. While GIS-based methods allow for integrating datasets that contain spatial variability, until recently datasets exhibiting spatio-temporal variability have been largely unmanageable, and researchers have been forced to simplify the complicated nature of their datasets by reducing or eliminating the spatial or temporal dimension. However, recent advances in space-time technology enable exposure scientists to more fully incorporate spatial and temporal variability into human exposure assessment. Whereas traditional GIS are based on spatial data structures--the "what, where" diad that inadequately displays changes through time, space-time technology is based on space-time data structures that enable characterization of the "what, where, when" triad needed for effective representation of data used to analyze health outcomes. Space-time technology allows the user to observe and quantify how geographies change with time, thereby enabling powerful exposure reconstruction procedures that are not possible through "space only" GIS. The continual expansion of space-time databases, coupled with the recognized need to incorporate human mobility in general and residential history in particular in environmental epidemiology, has highlighted the deficiencies of GIS-based software to visualize and process space-time information. This need is most pressing in retrospective studies where collection of individual biomarkers is unattainable or prohibitively expensive, and where models and software tools are required for exposure reconstruction. Advances in new space-time technology will profoundly improve our ability to reconstruct time-resolved individual exposures to environmental contaminants.

Bayesian Belief Networks (BBNs) are graphical models that incorporate probabilistic relationships among variables. Since their development in the late 1980s, they have increasingly been employed as tools for environmental modelling. Chapter 1 provides a summary of recent progress in using the technique, based on a review of relevant literature. A total of 50 publications were identified describing the application of BBNs to environmental problems, which were examined with respect to: (i) the development of model structures, (ii) identification of the probabilistic relationships between variables, (iii) the approaches used to test or validate models, (iv) sensitivity analysis, and (v) perceptions regarding the overall value of the approach. In the environmental sector, BBNs have to date primarily been used to examine management of natural resources, especially fisheries and water resources. The approach is widely considered to be useful, particularly in modelling domains characterized by high uncertainty, and where model parameterization is strongly dependent on expert knowledge. BBNs possess a number of features that make them particularly valuable in this context, especially where there is a need for models to support decision-making. However, many of the BBN models published to date are preliminary in nature, and there is a widespread lack of rigorous model testing, hindering a full evaluation of the approach. Current trends suggest that use of BBNs is likely to increase in future, extending to a wider range of domains, and involving increased integration with other modelling approaches and analysis of spatial data. However, increased emphasis on model testing is required if the approach is to fulfill its potential as a tool for environmental modelling.

Chapter 2 describes and demonstrates a methodology for the statistical retrieval of temperature and gas species concentration that uses the spectral radiance measured by new generation of high-resolution satellite-borne infrared sensors. These include, e.g., AIRS (Atmospheric InfraRed Sounder) on AQUA satellite and IASI (Infrared Atmospheric Sounding Interferometer) on the European Meteorological Operational satellite. These spectrometers are characterized by a wide band spectral coverage (645 to 2700 cm^{-1} or 3.7 to 15.5 μm) and a spectral sampling rate in the range 0.25 to 2 cm^{-1}. The performance of the retrieval scheme has been assessed on the basis of numerical exercises. Furthermore, examples of retrievals based on real spectra measured over sea surface are given to demonstrate the ability of the scheme to obtain accurate estimation of geophysical parameters. The problem of how many principal components to retain within the regression scheme has been addressed at a length and an original procedure is presented and discussed. Furthermore, the problem of statistical interdependency of retrieval and its vertical spatial resolution has been analyzed and a new index has been designed, which is capable to quantitatively deal with such an issue. The whole methodology has been derived for a generic signal-noise model and can therefore be used to design and implement retrieval algorithms also outside the specific area of high spectral resolution infrared observations

Insect infestations have been a major driving force of landscape change, leading to severe ecological and economic consequences. The southern pine beetle (SPB), *Dendroctonus frontalis* Zimmermann, is the most destructive insect to pine forests in the U.S. South. Chapter 3 probes the spatial and temporal patterns, particularly comovement and cyclical patterns, of SPB infestations at broad scales in the Southern United States. Cluster analysis in terms of comovement shows that SPB infestations in the region can be classified into three subregions: Alabama-Florida-Louisiana-Mississippi-Virginia, Georgia-Carolinas-Tennessee, and Arkansas-Texas. SPB infestation risk has increased over time in Florida, South Carolina, and Tennessee, but decreased in Alabama. The magnitude of bi-state comovements of SPB infestations is in general quite large whereas that of regionwide comovements is small, and comovments of small outbreaks are more pronounced than those of big ones. SPB infestations in North Carolina best resemble (synchronize with) the region's median, and thus it can be used as the region's reference for monitoring and forecasting. Though regionwide cyclical outbreak patterns are not detected, statistic evidence of cyclical outbreaks for some states and especially for the identified clusters/subregions is apparent; and the sinusoidal component is commonly concentrated over a range of low frequencies. These results will be of value for monitoring and mitigating SPB outbreaks in the region.

The objective of Chapter 4 is to provide a documented discussion on modelling of radionuclides interaction with suspended particulate matter in the marine environment. Aquatic environments occupy a major portion of the earth's surface and, therefore, an understanding of the radionuclide pathways is essential for radioecology purposes. Many radionuclides within such systems exhibit non-conservative behaviour, i.e., they undergo a change in the distribution between the particulate and dissolved phases, due to sorption reactions. The first reactive transport models for non-conservative radionuclides were based upon the equilibrium distribution coefficient, kd. However, recent models are based on kinetic rates of uptake/release of radionuclides between waters and the solid phases when short timescales are involved. Recently, much experimental and modelling effort has been focused on determining those factors which affect the kinetics and the final equilibrium conditions for the uptake of pollutants in aqueous suspensions under dynamic or static situations. Some of the obtained results appear to be either surprising or contradictory and introduce some uncertainty on which parameter values are most appropriate for environmental modelling.

The kinetic box models are widely used in modelling of radionuclides dispersion in aquatic systems, and are based on reversible reactions of constant coefficients. Indeed, several models have been proposed in the literature taking account of parallel, consecutive reactions and some combinations of both depending on the uptake experimental data. Research in this field of radioecology has been stimulated by the high level of public concern over health aspects of the presence of contaminant radionuclides in the environment, which has prompted a generally anti-nuclear posture. In this paper we present a review of the most relevant mathematical models developed so far and a brief revision of the historic developments of theories on surface-electrolyte interactions. Laboratory tracing experiments to study the uptake of ^{239}Pu, ^{241}Am, ^{133}Ba and ^{85}Sr in natural aqueous suspensions from several aquatic systems (reservoir, river, estuary and sea) are also presented to illustrate the application of box models covering a large number of environmental situations and different reaction channels.

The recent broadening of agricultural policy objectives and their related policy instruments has not been mirrored to the same extent in many agricultural sector models. For example, while the Common Agricultural Policy (CAP) of the European Union is becoming more concerned with rural development (second pillar of the CAP), most agricultural sector models are not able to handle the types of policy instruments covered by the second pillar accurately. Consequently, there is a considerable gap between the demand of policy makers and the supply provided by sector models. Chapter 5 aims at contributing to narrow this gap by presenting an extension of the agricultural sector model CAPRI (Common Agricultural

Policy Regional Impact Analysis) that enables a deeper regionalization of the model and which may prove useful in extending the relevance and reliability of indicators measuring agriculture's multifunctionality.

Chapter 6 summarises the main author's contributions in the past years over environmental modelling within the framework of the methodology called ACM ('Adaptive Control Methodology'). ACM combines the instruments of System Dynamics, multi-criteria analysis and control. Ex-ante control, i.e., prior to policy implementing is called 'control by structure' because it is achieved by developing control feed-back loops in the systemic structure under focus to influence the system evolution towards sustainable long-term paths. Ex-post control takes place during the post-implementation phase in which adjustments of the policy are necessary. ACM contributions in environmental modelling are briefly introduced, and a practical environmental case is presented on the way this technique can be used in real situations.

Distributed watershed models, such as the Soil and Water Assessment Tool (SWAT), have proven to be an important means for tackling problems related to nonpoint source pollution. These models require soil data as one of the minimal inputs because soil controls the runoff mechanism of a watershed. The U.S. Department of Agriculture Natural Resources Conservation Service generated two national soil databases, namely the State Soil Geographic (STATSGO) and Soil Survey Geographic (SSURGO) databases, which were designed to be used for river basin resource monitoring, planning, and management. STATSGO has a county level spatial resolution, whereas, SSURGO has a farm level resolution. These two databases have been widely used as the best alternatives for watersheds where site specific soil data are not available, which is a common case. However, in the literature, information is scarce regarding affects of using one of these two databases over another on modeling sediment processes. Thus, the objective of Chapter 7 was to evaluate affects of using STATSGO versus SSURGO as an input on SWAT simulated sediment yields at the outlet of, and sediment source areas within, a watershed. The evaluation was conducted in one North Dakota watershed and one Texas watershed. Obviously, these two watersheds have distinctly different climatic and hydrologic conditions. The results indicated that for each of the study watersheds, the predicted total sediment yields at the outlet using STATSGO might be comparable with the corresponding predicted values using SSURGO, the predicted source areas, however, could be very different. In addition, the prediction discrepancies as a result of using one dataset over another were larger in the Texas watershed. This might indicate that the simulation of a watershed with streamflows mainly generated from rainfall runoff is likely to be more sensitive to soil data resolution than the simulation of a watershed with streamflows predominantly generated by snowmelt runoff.

Because landslide generates a larger yearly loss of property than earthquake, flood, or windstorm, it is important to develop models that can measure the potential of landslide occurrence in an area. Landslide hazard models can be generally grouped into physically-based and statistical models. A physically-based model delineates areas prone to landsliding by analyzing the influence of surface topography on near-surface hydrologic response. It assumes that slope failures are caused by shallow subsurface flow convergence, increased soil saturation, and shear strength reduction. A statistical model predicts the likelihood of landslide occurrence by analyzing the relationship between past landslides and instability factors such as lithology, slope, curvature, aspect, elevation, land use, and drainage. Common statistical methods for landslide prediction include discriminant analysis and logistic regression. Model validation is the process of evaluating a landslide hazard model by comparing model predictions with observed landslides. Observed landslides are compiled from aerial photographs, satellite images, or from ground surveys. Chapter 8 covers landslide mapping, physically-based models, statistical models, model validation, and examples of models for a mountainous watershed in Taiwan.

Chapter 9 investigates the groundwater contamination by heavy metals of the industrial area of Jorf Lasfar, located on the Atlantic coast of El Jadida, Morocco. This paper describes groundwater mapping and data collection and management to be created. Water samples collected from wells near the area of the industrial installations immediately located on the coast for chemical analyses. Multivariate statistics and GIS techniques were applied to classify the elements, to facilitate interpretation of the spatial relationships among key environmental processes and to identify elements influenced by human activities.

The results show that the area in general is characterized by hard water and high salinity hazard, possibly due to its proximity and hydraulic connection with the sea. Fe, Mn, F, Cu, Pb, Ni, Cr and Cd were found to be the major contaminants in groundwater. The analysis of groundwater indicates contamination at various degrees. Spatial distribution modelling of element concentrations is produced to indicate contamination plumes from possible anthropogenic sources. It was observed that the groundwater in south-eastern of the Jorf Lasfar industrial area is contaminated due to industrial effluents in coast and predominant wind direction. Cluster analysis (CA) classified the elements into two groups: the first group being influenced by human activities, the second predominantly derived from natural sources.

EXPERT COMMENTARY

In: Environmental Modelling: New Research
Editor: Paul N. Findley, pp. 3-10
ISBN: 978-1-60692-034-3

Advances in Space-Time Technology for Assessing Human Exposure to Environmental Contaminants

***Jaymie R. Meliker*[1,2,*] *and Geoffrey M. Jacquez*[3, 4]**

[1]Graduate Program in Public Health, Department of Preventive Medicine, Stony Brook University;
[2]Consortium for Interdisciplinary Environmental Research, Stony Brook University;
[3]BioMedware, Inc.;
[4]Department of Environmental Health Sciences, University of Michigan

Abstract

In recent decades, digitally georeferenced data and Geographic Information Systems (GIS) have played a growing role in human exposure assessment for environmental epidemiologic studies. Despite the increased use of GIS in exposure assessment research, spatio-temporally varying datasets, such as daily activity spaces, residential histories, and time-varying maps of environmental contaminants are poorly characterized in the GIS environment. While GIS-based methods allow for integrating datasets that contain spatial variability, until recently datasets exhibiting spatio-temporal variability have been largely unmanageable, and researchers have been forced to simplify the complicated nature of their datasets by reducing or eliminating the spatial or temporal dimension. However, recent advances in space-time technology enable exposure scientists to more fully incorporate spatial and temporal variability into human exposure assessment. Whereas traditional GIS are based on spatial data structures--the "what, where" diad that inadequately displays changes through time, space-time technology is based on space-time data structures that enable characterization of the "what, where, when" triad needed for effective representation of data used to analyze health outcomes. Space-time technology allows the user to observe and quantify how geographies change with time, thereby enabling powerful exposure reconstruction procedures that are not possible through "space only" GIS. The continual expansion of space-time databases, coupled with the recognized need to incorporate human mobility in general and residential history in particular in environmental epidemiology, has highlighted the deficiencies of GIS-based software to visualize and process space-time information. This need

* E-mail address: jrmeliker@gmail.com

is most pressing in retrospective studies where collection of individual biomarkers is unattainable or prohibitively expensive, and where models and software tools are required for exposure reconstruction. Advances in new space-time technology will profoundly improve our ability to reconstruct time-resolved individual exposures to environmental contaminants.

Introduction

In recent years GIS have been adopted for reconstructing individual-level exposure to environmental contaminants in epidemiologic research [Beyea and Hatch, 1999; Nuckols et al., 2004; Ward et al., 2000]. The growing body of literature on this application most commonly includes studies that assess proximity of individuals to sources of environmental contaminants such as pesticide application on farms [Reynolds et al., 2005], landfill sites [O'Leary et al., 2004], and hazardous waste sites [Elliott et al., 2001; McNamee and Dolk, 2001]. On occasion, investigators implement environmental fate and transport models to improve the exposure assessment [Reif et al., 2003]. These studies, however, rely almost exclusively on home residence at time of interview/diagnosis as the spatial location of the individual. Yet individuals are mobile and frequently change residences, and this residential mobility information is crucial for analyses of chronic diseases with relatively long induction periods, such as cancer.

Failure to include space-time mobility in GIS-based exposure assessments is a consequence, at least in part, of the a-temporal nature of GIS. GIS operate within a static world-view which is largely incapable of representing temporal change [Goodchild, 2000]. GIS are best suited to "snapshots" of static systems [Hornsby and Egenhofer, 2000] which hinders the mapping, representation, and analysis of dynamic health, socioeconomic, and environmental information for mobile populations. In the few GIS-based exposure assessments where residential histories are included, researchers create numerous maps, or snapshots, in specified time intervals (usually annual) to assess exposure as individuals move through time [Aschengrau et al., 1996; Bellander et al., 2001; Bonner et al., 2005; Brody et al., 2002; Nyberg et al., 2000, Stellman et al., 2003; Swartz et al., 2003]. Assembling these snapshots, however, is plagued by a host of problems: (a) it is dataset-intensive, requiring a unique database and map for each time slice; (b) it is labor-intensive and as such has the potential to produce critical errors during data manipulation; (c) information about change is not available in the interval between two consecutive snapshots; (d) it is so time consuming that it typically only enables one attempt at exposure reconstruction--it does not allow for improvements in the underlying models of environmental contamination for refining and recalculating exposure in an iterative manner; and (e) dozens-to-hundreds of maps need to be created to calculate time-resolved exposure using different temporal orientations or "measures of time", such as participants' age, calendar year, and years prior to diagnosis/interview. Perhaps most problematic is the fact that the snap-shot approach implicitly involves time-averaging and the implicit assignment of values of the exposure metric over the duration of the defined snapshot. This ignores or underestimates temporal variability inherent in the exposure metric and can result in biased exposure estimates. In short, the "snap shot" approach is both difficult to implement and yields misleading results. As a result of these challenges and limitations, when conducting GIS-based exposure assessments most researchers choose to disregard residential histories, thereby implicitly assuming that individuals are immobile, and that the induction period between causative exposures and health events (e.g. diagnosis, death) is negligible [Jacquez, 2004]. These strong assumptions are typically invalid, and the resulting exposure estimates are of dubious validity and unknown accuracy. The availability of software for displaying and integrating information across space-time maps that treat time as continuous, rather than discrete, could profoundly improve our ability to reconstruct time-resolved individual exposures [AvRuskin et al., 2004; Meliker et al., 2005].

We are not alone in identifying these weaknesses in GIS-based exposure assessments. In a recent review of epidemiological methods for evaluating geographic exposures and hazards, Mather et al. [2004] lamented the scarcity of methods that account for residential histories of cases and controls. At a meeting of this nation's experts on the spatial analysis of cancer data, the need to account for latency and human mobility in cancer studies was recognized as the second most pressing issue [Pickle et al., 2005]. As Pickle and colleagues [2006] summarize from that meeting, "...few tools and methods can be applied to both space and time together... ...a data representation problem needs to be solved: how to store and retrieve integrated space-time data consisting of multiple sets of data from fundamentally different space-time frames." This problem exists because, despite modern computer technologies for storing and managing temporal or spatial datasets, surprisingly few tools are available for working with space-time datasets [Beaubroef and Breckenridge, 2000; Dragicevic and Marceau, 2000]. Without tools for visualizing and analyzing space-time datasets, participant mobility, changes in contaminant concentrations, and other forms of space-time variability are inadequately incorporated into human exposure assessments.

Space-Time Software Systems

In the past few years several new software packages have emerged for working with space-time data. To our knowledge, currently available space-time visualization software include GeoTime, TimeMap, Space Time Toolkit, Google Earth, and ArcGIS 9.2; however, analysis capabilities necessary for exposure assessment are highly limited in these software. For example, the capacity to join data between space-time datasets of mobility and environmental contaminants is not available in these software. Another software package that provides both space-time visualization and some analysis capability has yet to be released (STNexus from GeoVista), and we are unsure of its ability to join space-time datasets, either. TerraSeer's Space-Time Intelligence System or STIS software, initially released in 2004, represents the only package that we know of which allows space-time visualization, data management, and analysis, and is capable of joining information between space-time datasets for exposure assessment. STIS technology was developed over several years with funding from the National Institutes of Health, with the specific objective of developing data structures, visualization, and analysis approaches for assessing health-environment relationships using time-dynamic data. While traditional GIS are based on spatial data structures, STIS is based on space-time data structures that are essential for the effective representation of data used to analyze health outcomes. STIS allows the user to observe and quantify how geographies change with time, thereby enabling powerful exposure reconstruction procedures that are not possible through "space only" GIS. It is ideally suited to the representation, visualization and reconstruction of human exposures to environmental contaminants.

Space-Time Datasets for Exposure Assessment

Space-time datasets relevant for exposure assessment are increasingly available. These resources include mobility histories from self-report, centralized database, and location-enabled devices (e.g., global positioning system (GPS)). They also include environmental contamination histories from temporally-rich monitoring networks, space-time models of monitored data, and remotely sensed data from satellite imagery and airborne platforms. These datasets will prove immeasurably valuable for reconstructing historic exposures to environmental contaminants.

Location-enabled devices track the locations (e.g. *x, y* coordinates and sometimes altitude) of individuals and assets (e.g. appliances such as Xerox machines in buildings, trucks, packages, etc.). This may be accomplished using GPS chips in cellular phones, RFID (radio frequency identification) tags on packages and shipments, and sensing technologies within and around buildings that employ ultrasound, WI-FI, UWB (Ultra-wide bandwidth) and related sensing technologies. The latter typically involve a tag being worn on the asset or individual, a sensor that determines the position of the tag in geographic space down to 0.5m resolution, and that transmits this location data at temporal sampling intervals up to several times per minute over a local network for storage and processing.

Imagery from satellites (e.g., Landsat) and air-borne platforms are finding increased use in exposure assessment. Applications include exposures to agricultural herbicides and pesticides based on proximity of homes to agricultural fields, where the field type is classified from satellite imagery. Knowledge of the crop type may then be coupled with information on agricultural practices such as the amounts of herbicide and pesticide use for that crop type throughout the year. GIS procedures may then be used to rank exposures at residential locations based on proximity to fields. As imaging technologies have advanced they have increased temporal (return visits during flyovers), spatial (pixel size), and spectral resolution (width and number of individual bands that report the amount of reflectance in specific bandwidths). Whereas Landsat could resolve individual pixels on the order of 30 meters using 6 bands, new hyperspectral technologies can resolve pixels of 0.5 meters with 160 or more bands. This makes possible the classification of land-based features with greater precision and accuracy.

Extensive monitoring networks exist for sensing environmental parameters such as air pollutants (e.g. ozone, NOx, SOx, etc.), temperature, rainfall and acidity, stream flows, snow pack and a host of other variables. Efforts are currently underway involving governments around the world to create a "system of systems" for monitoring the earth. In February, 2005, member countries of the Group on Earth Observations agreed to a 10-year implementation plan for a Global Earth Observation System of Systems (GEOSS). The GEOSS project holds the potential to advance exposure assessment by producing and managing information in a way that makes it readily accessible to a diversity of users in common formats. At this writing it is not entirely clear how GEOSS data may best be used for exposure assessment. A pressing practical issue is the integration of data from a diversity of sources at different spatial scales, from points (e.g. monitoring station locations), to polygons (e.g. the extent of municipal water supply districts) to rasters (e.g. Landsat Imagery); at different spatial resolutions (from submeter to several kilometers); and employing different temporal sampling frequencies (from essentially static – one observation – to several times per second, as occurs for some air sampling monitoring devices). And how might such a diversity of data be combined with data on individual-level mobility to derive time-dependent exposure measures? This is a significant research challenge with high potential payoff.

Example Applications of Space-Time Exposure Reconstruction

To our knowledge, at time of writing, space-time software has only been applied one time for reconstructing human exposure. In that application, individual lifetime exposure to arsenic in drinking water was assessed among participants in a population-based bladder cancer case-control study [Meliker et al., 2007]. Information collected included residential mobility, changes in individual risk behaviors such as water and food consumption and smoking, occupational history, estimates of arsenic in past private wells, the changing geography of municipal water supply districts, as well as the space-time variability of arsenic concentrations in municipal drinking water sources. Arsenic concentrations were automatically assigned to each individual's home drinking water by successively looping through each person's residential history and assigning an arsenic value for each space-time interval. The estimate of arsenic concentration in home drinking water came from a geostatistically-derived raster image for private well arsenic concentration [Goovaerts et al., 2005], and space-time estimates of arsenic in municipal water supplies. Every participant was assigned an arsenic concentration for each unique water supply at a residence. This resulted in an estimate of arsenic concentration in home drinking water for each moment in a person's life. By automating this method, it allowed effortless recalculation of individual lifetime exposure (1) whenever an underlying dataset was improved, and (2) for exploring alternate temporal orientations (e.g., age, calendar year, and years prior to diagnosis/interview).

There are numerous additional opportunities for space-time exposure reconstruction. For example, individuals change locations over the course of a day and are exposed to different concentrations of air pollutants depending on when and where they are located. Given datasets of individual daily mobility patterns and hourly concentrations of air pollution, a space-time join function could assign the value of an air pollutant concentration to each person at every location for the appropriate temporal interval. In another example taking advantage of land cover classifications (e.g., available from Landsat satellite imagery) and residential histories, the proximity of each individual to farmland could be calculated, again, at every location for the appropriate time interval. Estimates of air pollution concentrations and proximity to farmland are common measures of exposure in environmental epidemiologic studies [Bellander et al., 2001; Swartz et al., 2003]; however, prior to development of STIS technology, temporal dimensions of these geographic exposure assessments were simplified by only relying on time-slices.

Conclusion

Individuals and environmental contaminants change locations in time, and therefore tools that integrate space-time maps are necessary to more reliably calculate individual exposure. Space-time technology has recently been developed and has the potential to advance the assessment of individual-level exposure to environmental contaminants. Space-time exposure reconstruction explicitly characterizes the timing and ordering of exposure, accounting for temporal changes in the magnitude and locations of exposure. By efficiently and easily incorporating space-time variability, space-time technology enables flexibility in the assessment of temporally detailed individual-level exposure estimates. This flexibility in integrating information across space-time datasets allows the user to calculate a series of alternative exposure metrics and to refine the exposure estimates as the underlying models of environmental contamination are improved. This type of iterative process is impractical with traditional GIS because of the time and labor involved in generating an exposure estimate. These improved estimates should be valuable for exposure assessment in environmental and cancer epidemiology studies and for risk assessment, surpassing GIS-based efforts that are less reliable and less accurate because they ignore the spatio-temporal variability inherent in many datasets. Future work needs to (1) describe additional applications of space-time technology for exposure assessment, (2) examine the magnitude of the improvement in exposure assessment that comes from using space-time technology, and (3) investigate whether macro- (e.g. residential histories) and micro-scale (e.g. daily activity patterns) data can be easily integrated to support exposure assessment for individuals. The time geographic approach constitutes a valuable tool for advancing traditional means of exposure reconstruction.

References

Aschengrau A, Ozonoff D, Coogan P, Vezina R, Heeren T, Zhang Y (1996) Cancer risk and residential proximity to cranberry cultivation in Massachusetts. *Am J Public Health* **86**:1289-96.

Avruskin GA, Jacquez GM, Meliker JR, Slotnick MJ, Kaufmann AM, Nriagu JO (2004) Visualization and exploratory analysis of epidemiologic data using a novel space time information system. *Int J Health Geogr* **3**:26.

Beaubroef T, Breckenridge J. 2000. Real-world issues and applications for real-time geographic information systems (RT-GIS). *J Navigation* **53**:124-131.

Bellander T, Berglin N, Gustavsson P, Jonson T, Nyberg F, Pershagen G, Jarup L (2001) Using geographic information systems to assess individual historical exposure to air pollution from traffic and house heating in Stockholm. *Environ Health Perspect* **109**:633-639.

Beyea J, Hatch M (1999) Geographic exposure modeling: a valuable extension of geographic information systems for use in environmental epidemiology. *Environ Health Perspect* **107**(suppl 1):181–190.

Bonner MR, Han D, Nie J, Rogerson P, Vena JE, Muti P, Trevisan M, Edge SB, Freudenheim JL (2005) Breast cancer risk and exposure in early life to polycyclic aromatic hydrocarbons using total suspended particulates as a proxy measure. *Cancer Epidem Biomar* **14**:53-60.

Brody JG, Vorhees DJ, Melly SJ, Swedis SR, Drivas PJ, Rudel RA (2002) Using GIS and historical records to reconstruct residential exposure to large-scale pesticide application. *J Expo Anal Env Epid* **12**:64-80.

Dragicevic S, Marceau DJ (2000) A fuzzy set approach for modeling time in GIS. *Int J Geog Inf Sci* **14**:225-245.

Elliott P, Briggs D, Morris S, de Hoogh C, Hurt C, Jensen TK, Maitland I, Richardson S, Wakefield J, Jarup L (2001) Risk of adverse birth outcomes in populations living near landfill sites. *BMJ* **323**:363-368.

Goodchild M (2000) GIS and Transportation: Status and Challenges. *GeoInformatica* **4**:127-139.

Goovaerts P, AvRuskin G, Meliker J, Slotnick M, Jacquez GM, Nriagu J (2005) Geostatistical modeling of the spatial variability of arsenic in groundwater of Southeast Michigan. *Water Resour Res* **41**(7):W07013 10.1029.

Hornsby K, Egenhofer M (2000) Identity-based change: a foundation for spatio-temporal knowledge representation. *Int J Geog Inf Sci* **14**:207-224.

Jacquez GM (2004) Current practices in the spatial analysis of cancer: flies in the ointment. *Int J Health Geogr* **3**:22.

Mather FJ, Whited LC, Langlois EC, Shorter CF, Swalm CM, Shaffer JG, Harley WR (2004) Statistical methods for linking health, exposure and hazards. *Environ Health Perspect* **112**:1440-1445.

McNamee R, Dolk H (2001) Does exposure to landfill waste harm the fetus? *BMJ* **323**:351-352.

Meliker JR, Slotnick MJ, AvRuskin GA, Kaufmann A, Jacquez GM, Nriagu JO (2005) Improving exposure assessment in environmental epidemiology: Application of spatio-temporal visualization tools. *Journal of Geographical Systems* **7**:49-66.

Meliker JR, Slotnick MJ, AvRuskin GA, Kaufmann A, Fedewa SA, Goovaerts P, Jacquez GM, Nriagu JO (2007) Individual lifetime exposure to inorganic arsenic using a Space-Time Information System. *Int Arch Occ Env Health* **80**:184-197.

Nuckols JR, Ward MH, Jarup L (2004) Using Geographic Information Systems for Exposure Assessment in Environmental Epidemiology Studies. *Environ Health Perspect* **112**:1007-1015.

Nyberg F, Gustavsson P, Jarup L, Bellander T, Berglind N, Jakobsson R, Pershagen G (2000) Urban air pollution and lung cancer in Stockholm. *Epidemiology* **11**:487-495.

O'Leary ES, Vena JE, Freudenheim JL, Brasure J (2004) Pesticide Exposure and Risk of Breast Cancer: a Nested Case-Control Study of Residentially Stable Women Living on Long Island. *Environ Res* **94**:134-144.

Pickle L, Waller L, Lawson AB (2005) Current Practices in cancer spatial data analysis: a call for guidance *Int J Health Geogr* **4**:3.

Pickle LW, Szczur M, Lewis DR, Stinchcomb DG (2006) The crossroads of GIS and health information: a workshop on developing a research agenda to improve cancer control. *Int J Health Geog* **5**:51.

Reif JS, Burch JB, Nuckols JR, Metzger L, Ellington D, Anger WK (2003) Neurobehavioral Effects of Exposure to Trichloroethylene Through a Municipal Water Supply. *Environ Res* **93**:248-258.

Reynolds P, Hurley SE, Gunier RB, Yerabati S, Quach T, Hertz A (2005) Residential proximity to agricultural pesticide use and incidence of breast cancer in California, 1988-1997. *Environ Health Perspect* **113**:993-1000.

Stellman JM, Stellman SD, Weber T, Tomasallo C, Stellman AB, Christian R (2003) A Geographic Information System for Characterizing Exposure to Agent Orange and Other Herbicides in Vietnam. *Environ Health Perspect* **111**:321-328.

Swartz CH, Rudel RA, Kachajian JR, Brody JG (2003) Historical Reconstruction of Wastewater and Land Use Impacts to Groundwater Used for Public Drinking Water: Exposure Assessment Using Chemical Data and GIS. *J Expo Anal Env Epid* **13**:403-416.

Ward MH, Nuckols JR, Weigel SJ, Maxwell SK, Cantor KP, Miller RS (2000). Identifying populations potentially exposed to agricultural pesticides using remote sensing and a geographic information system. *Environ Health Perspect* **108**:5-12.

RESEARCH AND REVIEW STUDIES

In: Environmental Modelling: New Research
Editor: Paul N. Findley, pp. 13-50

ISBN: 978-1-60692-034-3

Chapter 1

BAYESIAN BELIEF NETWORKS IN ENVIRONMENTAL MODELLING: A REVIEW OF RECENT PROGRESS

Adrian C. Newton

Centre for Conservation Ecology and Environmental Change, School of Conservation Sciences, Bournemouth University, Talbot Campus, Poole, Dorset BH12 5BB, UK

Abstract

Bayesian Belief Networks (BBNs) are graphical models that incorporate probabilistic relationships among variables. Since their development in the late 1980s, they have increasingly been employed as tools for environmental modelling. This chapter provides a summary of recent progress in using the technique, based on a review of relevant literature. A total of 50 publications were identified describing the application of BBNs to environmental problems, which were examined with respect to: (i) the development of model structures, (ii) identification of the probabilistic relationships between variables, (iii) the approaches used to test or validate models, (iv) sensitivity analysis, and (v) perceptions regarding the overall value of the approach. In the environmental sector, BBNs have to date primarily been used to examine management of natural resources, especially fisheries and water resources. The approach is widely considered to be useful, particularly in modelling domains characterized by high uncertainty, and where model parameterization is strongly dependent on expert knowledge. BBNs possess a number of features that make them particularly valuable in this context, especially where there is a need for models to support decision-making. However, many of the BBN models published to date are preliminary in nature, and there is a widespread lack of rigorous model testing, hindering a full evaluation of the approach. Current trends suggest that use of BBNs is likely to increase in future, extending to a wider range of domains, and involving increased integration with other modelling approaches and analysis of spatial data. However, increased emphasis on model testing is required if the approach is to fulfill its potential as a tool for environmental modelling.

Introduction

A Bayesian Belief Network (BBN, sometimes referred to as a Bayesian network or Belief net) may be defined as a graphical model that incorporates probabilistic relationships among variables of interest (Heckerman 1997). The term graphical model is used because the BBN can be represented in the form of a network diagram, to provide a visual representation of the components and dependencies of a domain. Bayesian networks evolved in the late 1980s through developments of theory developed for graphical models, particularly through the work of Pearl (1986, 1988, 1995). This research established BBNs at the interface between statistics, applied artificial intelligence and expert system development. BBNs may be used as tools for building predictive models, for structuring knowledge or beliefs relating to a domain, or for supporting decision making.

It was recognized at an early stage that Bayesian networks have particular potential for the development of expert systems (Lauritzen and Spiegelhalter 1988), and early research on this theme was undertaken primarily in the field of medical diagnosis (Spiegelhalter *et al.* 1993). The application of BBNs to environmental modelling and natural resource management problems began in the early 1990s, with Varis *et al.* (1990) and Haas (1991) providing early examples.

In recent years, interest in the use of BBNs for environmental modelling has grown steadily, and examples are now available illustrating application of the method to a wide variety of different problems. The growing interest in using BBNs as a tool for environmental modelling is linked to the increasing acceptance of Bayesian approaches in ecology and environmental science (Ellison 1996, Clark 2005). However, it is important to differentiate BBNs from other approaches involving Bayesian analysis. Typically, Bayesian statistics are used to estimate parameter values when the stochastic component of a model is represented by one or more continuous probability density functions (Clark 2005). In common with other Bayesian approaches, BBNs can be considered as tools for updating existing (*a priori*) information using probabilities as a measure of uncertainty (Haapasaari *et al.* 2007). However, BBNs are characterized by the graphical representation of the model as directed acyclic graph (DAG). This review explicitly focuses on BBNs and their use in environmental modelling, rather than the broader use of Bayesian approaches to statistical analysis.

This chapter presents a review of recent progress in the use of BBNs in environmental modelling, by providing a summary of the results of recent research investigations. This was achieved by undertaking a literature search to identify relevant scientific publications, which were then summarized with reference to a number of key themes relating to use of BBNs as a modelling approach. Specifically, the review focuses on: (i) the development of model structures, (ii) identification of the probabilistic relationships between variables, (iii) the approaches used to test or validate models, (iv) sensitivity analysis, and (v) the conclusions reached by the authors concerned regarding the value of the method. This chapter provides the first systematic overview of recent experience in using BBNs in environmental modelling, and is designed to provide an evaluation of the strengths and weaknesses of the technique, based on recent evidence. The review provides a summary of the research experiences achieved to date, and concludes with some speculation regarding how the approach might develop in future, based on current trends.

Description of a BBN

In a Bayesian network diagram, variables, data and parameters are represented by different shapes (such as ellipses and rectangles), which are connected by arrows to indicate conditional dependencies (Figure 1). The ellipses representing variables are referred to as *nodes*, whereas the arrows are referred to formally as *directed links* (or alternatively as *edges* or *arcs*) (Jensen 2001). A link between two nodes, from node A (parent node) to node B (child node), indicates that A and B are functionally related, or that A and B are statistically correlated. Each child node (i.e. a node linked to one or more parents) contains a conditional probability table (CPT). The CPT gives the conditional probability for the node being in a specific state given the configuration of the states of its parent nodes. For variables without parents, an unconditional (or marginal) probability distribution is defined (unconditional probabilities being those that are independent of other variables). When networks are compiled, Bayes' theorem is applied according to the values in the CPT, so that changes in the probability distribution for the states at node A are reflected in changes in the probability distribution for the states at node B (Jensen 2001). A BBN can be explored by changing the states of the nodes (or variables) incorporated within the model. When the state of a node is known, it is said to be *instantiated* (Jensen 2001). Once a node has been instantiated, then this will influence the probabilities associated with the states of other nodes to which it is linked, according to the values in the CPTs. When a BBN is compiled, results are therefore presented in the form of probability distributions rather than single values.

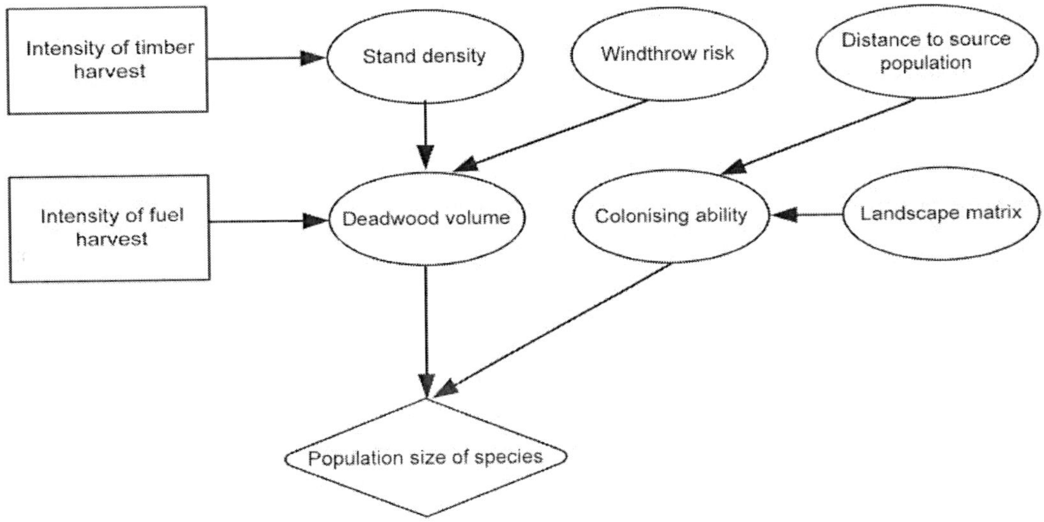

Figure 1. Illustration of a simple Bayesian Belief Network. The example depicted is an influence diagram constructed to inform conservation of a threatened fungus species, which depends on deadwood as habitat (adapted from Newton 2007). Different shapes (or nodes) on the diagram illustrate different kinds of variable. A rectangular node depicts a *decision node*, which the decision-maker can control directly (in this case, harvesting of trees for timber and deadwood for fuelwood). The elliptical nodes represent variables that the decision-maker cannot control directly. The rhombus-shaped node is a *utility node*, which represents the desired outcome, in this case a viable population of the species concerned. The arrows in the diagram represent influences among the variables.

The network of a BBN is referred to as a directed acyclic graph (DAG), as graph links have directions but no directed cycles are permitted. In other words, it is not permitted to form a closed loop in a BBN. BBN models are therefore unable to explicitly represent system feedbacks (Borsuk *et al.* 2004), in marked contrast to approaches typically used for modelling ecological systems, such as compartment-flow models (Grant and Swannack 2008). However, in contrast to most other modelling approaches, propagations can be made through the links in either direction of the network, enabling the model to be explored in reverse: the BBN can be used to infer the most likely set of conditions for a given outcome (McCann *et al.* 2006). Many authors consider the links between nodes to represent causal relationships (Pearl 1986), but this does not necessarily have to be the case. Strictly, the directed links are representations of conditional dependence, and therefore may more accurately be described as representing influence rather than causality (Lauritzen and Spiegelhalter 1988).

BBNs differ from most other approaches to environmental modelling by exclusively using probabilistic, rather than deterministic, expressions to describe the relationships among variables, a feature that is particularly useful in the context of risk assessment and for supporting decision making (Borsuk *et al.* 2004). A BBN represents a description of the probabilistic relationships among the domain's variables, enabling the joint (or total) probability distribution of all variables to be divided into a series of conditional and unconditional distributions (Borsuk *et al.* 2004). The graphical nature of the network facilitates visualization of such relationships, and as recognized by many authors, its visual nature can foster communication among scientists, decision-makers and other stakeholders regarding the relationships between variables in a particular domain.

Bromley *et al.* (2005) stressed that BBN models do not necessarily replace existing environmental, economic or social models; instead it is possible to take the outputs from such models and incorporate them in BBNs, by converting them into probability distributions, an approach that appears to becoming increasingly popular. BBNs can therefore be seen as tools for integrating different kinds of knowledge or evidence, including model outputs and quantitative data, together with subjective information such as expert knowledge (Borsuk *et al.* 2004). This ability to integrate both qualitative and quantitative information is widely considered to be one of the main assets of BBNs (Newton *et al.* 2006), and is particularly welcomed by researchers investigating environmental management, who may often be required to integrate socio-economic with biophysical information.

Another key feature of BBNs is that they provide a tool for reasoning under uncertainty (Jensen 2001), which may be defined as a lack of information or knowledge (McCann *et al.* 2006). Various different types of uncertainty can be distinguished, including an imperfect understanding of the domain, incomplete knowledge of the state of the domain, randomness in the mechanisms governing the behaviour of the domain, or any combination of these (Ma *et al.* 2004, Bromley *et al.* 2005, Regan *et al.* 2002). BBNs can model uncertainty in terms of the probability distribution associated with the states of the variables; the degree of uncertainty is reflected in the uniformity of the probability distributions obtained, with higher uncertainty associated with more even distributions (Ma *et al.* 2004). Environmental problems are typically associated with high uncertainty, because of a lack of appropriate information or understanding about any given domain, and how this may change over time (McCann *et al.* 2006). BBNs may therefore be a particularly valuable approach for environmental modelling, especially where environmental management decisions need to be made.

Bromley *et al.* (2005) highlight the fact that BBNs are especially useful to support decision making in relating to natural resource management, because typically a large number of interlinked factors need to be taken into consideration, and data are often scarce and uncertain. BBNs possess a number of features designed to support their use as decision-support tools, including the potential inclusion of *decision nodes* and *utility nodes* (Figure 1).

Decision nodes represent decisions made by a user of the model; they have no CPTs associated with them. A utility node represents some measure that can be used to assess the success or failure of a decision, such as economic cost or environmental value (Sadoddin *et al.* 2005, Nyberg *et al.* 2006). Utility nodes are associated with a utility table, which specifies the utility of each configuration of the parents of the utility node (which may be influenced by a decision, through a link with a decision node). A BBN that includes utility and decision nodes is often referred to as an *influence diagram* (although some authors use this term more widely, to apply to graphical models more generally). Influence diagrams are closely related to decision trees, but differ in their graphical construction (Kuikka *et al.* 1999, McCann *et al.* 2006). Once parameterized, such a model can be explored to identify the choice in each decision node that minimizes the costs or maximizes the benefits or values considered (Nyberg *et al.* 2006). The expected values for each choice in the decision nodes are displayed by the BBN, produced by combining utilities and their associated probabilities (Nyberg *et al.* 2006).

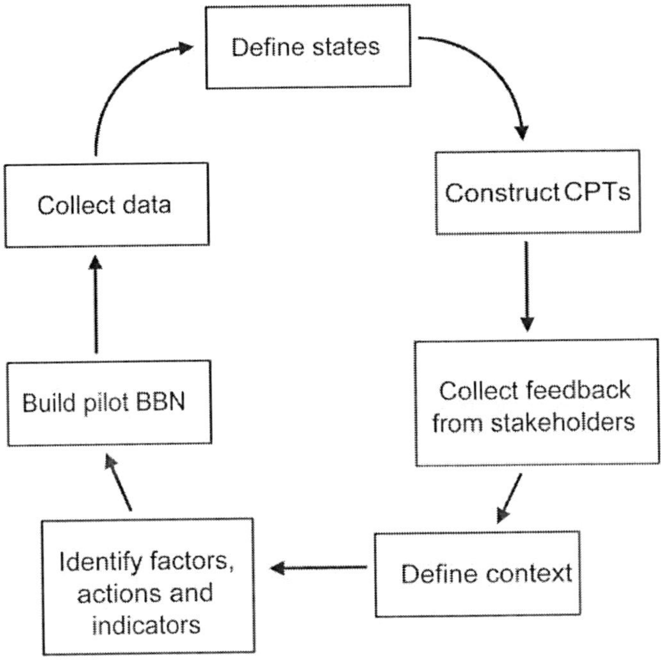

Figure 2. A proposed procedure for involving stakeholder consultation in BBN development, involving seven steps (adapted from Henriksen *et al.* 2007). Step 1: Define context: the scope of the domain to be modeled is defined, including physical and socio-economic boundaries. Step 2: Identify factors, actions and indicators: a list of concerns identified by the stakeholders is compiled, and actions to be taken and important indicators are identified. Step 3: Build pilot BBN, involving construction of a network diagram. Step 4: Collect data, and analyse them to inform model development. Step 5: Define states of the variables, with input from stakeholders. Step 6: Construct CPTs, using stakeholder meetings, together with input from available data, models and experts. Step 7: Collect feedback from stakeholders on the final network.

The practical development of a BBN begins with construction of a network diagram representing the domain of interest, which is often based on expert knowledge or the judgement of researchers, and is sometimes refined through a process of stakeholder consultation (Figures 2 and 3). The creation of a network diagram can itself be a useful way of eliciting information from experts and structuring the information available. A number of different model structures may be explored, with the aim of identifying the structure that best captures the logical relationships between the variables being considered.

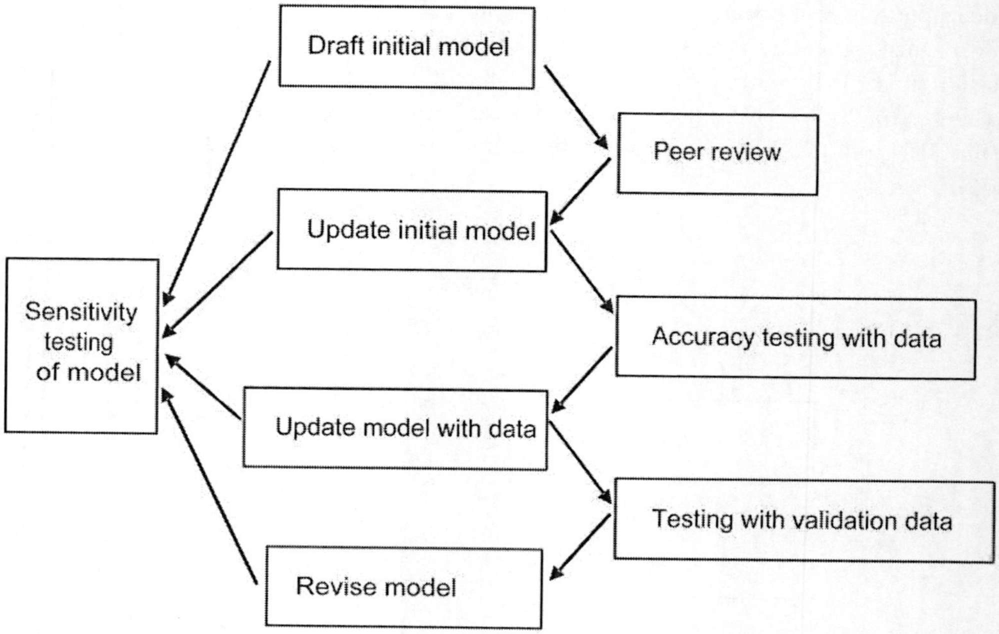

Figure 3. Illustration of the process by which a BBN prediction model may be developed (adapted from Marcot 2006). The process involves construction of a series of versions of the model, which are each revised in the light of comments from expert consultation and from the results of testing with data. Sensitivity analysis is also performed at each step of the process, and results from these analyses are also used to revise the model.

Once the structure has been defined, the main challenge is to complete the CPTs that are associated with each node. These define the probabilistic relationships between the nodes connected by directed links. As summarized by Kocabas and Dragicevic (2006), the elements of a BBN are therefore: (i) the variables represented by nodes and the set of directed links connecting them, producing the overall structure of the DAG; (ii) the states of each of the variables, and (iii) the CPTs.

Overall, the creation of a BBN may therefore be considered as comprising three stages (Jensen 2001, Martín de Santa Olalla *et al.* 2005):

(1) *Identification of the variables* that are relevant to the domain that is to be modeled.

(2) *Establishment of the relationships between the variables,* which are represented by the arrows linking the variables in a network diagram.

(3) *Assignment of states and probabilities to each variable,* when the potential values (or states) of the variables are defined, together with the probabilities associated with each of them, as entered in the CPTs associated with each node.

Further details of BBNs are provided by Castillo *et al.* (1997), Fenton and Neil (2007), Jensen (1996, 2001), Olson *et al.* (1990), Reckhow (2003), Spiegelhalter *et al.* (1993), and Varis (1997). Recent reviews of the use of BBNs in natural resource management are provided by McCann *et al.* (2006), Nyberg *et al.* (2006), Varis (1998) and Varis and Kuikka (1999), and guidelines for their use in this context are provided by Cain (2001) and Marcot *et al.* (2006).

Literature Search Methods

A literature search was conducted using Web of Knowledge (http://www.isiwebofknowledge.com/) and Google (http://www.google.co.uk) search engines, using the search term 'Bayesian network', with a publication date up to and including 2008. Those publications involving the application of BBNs to environmental science and management, based on inspection of the publication title and abstract, were shortlisted for further examination of the complete text. The reference lists of shortlisted publications were also consulted, in order to identify additional publications. The search focused primarily on publications in international, refereed scientific journals; those available only as unpublished 'grey' literature that were difficult to access were excluded from the review. A broad, inclusive definition of 'environmental science and management' was implemented when selecting publications for inspection. Only those publications that provided an actual example of a BBN were included in the survey; those that merely described the technique were excluded.

The final list of selected publications that was produced as a result of this process cannot considered to be either unbiased or comprehensive. For example, not all scientific journals will necessarily have been included by these search engines, and some relevant publications may not have been detected using these search terms. In addition, publications not written in English would have stood little chance of detection. It is likely that many other relevant examples exist within the 'grey' literature, including those presented in postgraduate theses (e.g. Taylor 2003), which were excluded from this review. Despite these caveats, it is likely that the shortlisted publications provide a representative overview of recent research in this field.

Each shortlisted publication was examined in detail, and information extracted regarding: (i) the domain or scope of the investigation, (ii) how the structure of the BBN was produced, (iii) how the CPTs were completed, (iv) how the model was validated (or verified), (v) whether or not a sensitivity analysis was performed, and (vi) the conclusions reached with respect to the value of the method. In extracting the information in this way, some judgement was often required to interpret the information provided, as in many cases, the methods used were not fully described in the text. It should therefore be remembered that the information presented here is based on my interpretation of what the authors presented, and the original publications should be consulted for further details.

In total, 50 published studies were included in the review. A summary of the information extracted from this literature is presented on Table 1. The following sections consider each of the issues summarized on the table in greater depth, with reference to specific examples.

Domains

Each published study included in the review was assigned to a broad category of domain, to provide an indication of the distribution of studies among environmental sectors (Figure 4).

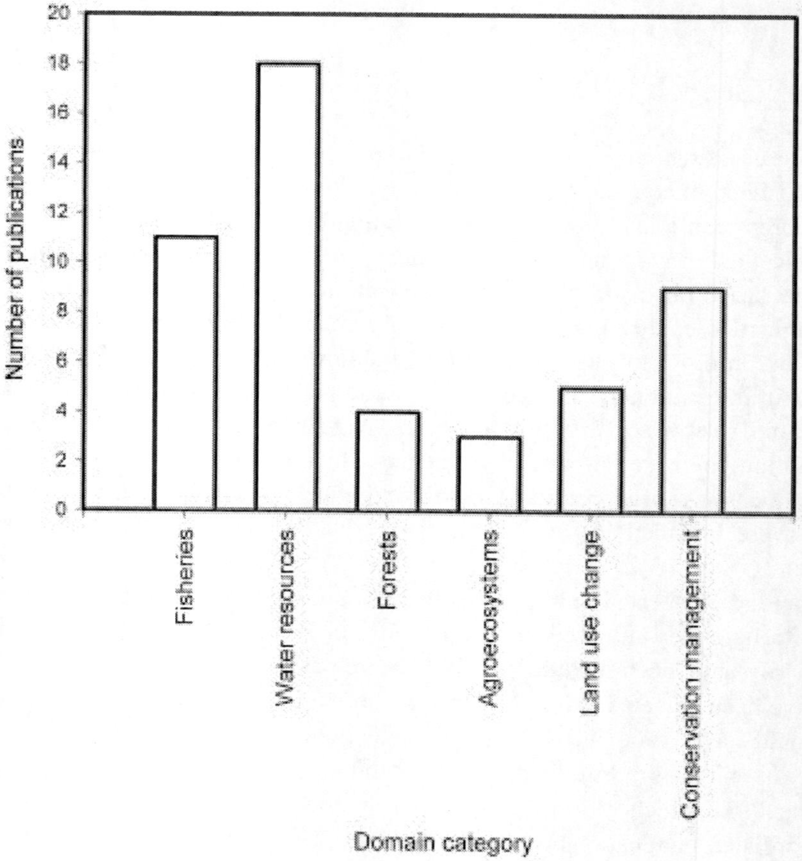

Figure 4. Distribution of publications included in the literature review (see Table 1) among different broad categories of domain. For the purposes of this classification, each publication was assigned to one category only, selecting the category that best described the focus of the investigation in question, namely fisheries (including fisheries management and analysis of the population dynamics of fish species), water resources management (including pollution impacts, groundwater quality and waste water management), forests (including forest management and sustainable use), agroecosystems, land use change (including land planning and ecosystem mapping), and conservation management (principally of mammals and birds).

This classification is necessarily very approximate, as the classes adopted are not mutually exclusive – for example, studies focusing on catchment management could arguably be relevant to all of the categories considered here. However, the results indicate that two subject areas have been the focus of particular research activity: management of fish populations, and water resource management. Investigations within these categories addressed a variety of different problems; for example, publications relating to water resource management encompassed studies of pollution, catchment management, water quality and wastewater treatment. Other domains were less well represented; conservation management (including habitat suitability analysis) accounted for nine investigations, whereas other categories were each represented by five publications or fewer (Figure 4). Perhaps surprisingly, only two investigations were explicitly designed to explore the potential impacts of climate change. In addition, few studies addressed spatial planning or management; linkage of BBN models with geographical information systems (GIS) still appears to be its infancy.

Network Structure

The advantages of a relatively simple network structure are highlighted by Marcot (2006) and Marcot et al. (2006), who presented the following recommendations:

- Specify no more states per node than are necessary,
- Keep the number of parent nodes to four or fewer (or even two or fewer; Bromley *et al.* 2005), to ensure that the CPTs are manageable
- Build models broad rather than deep; in other words, try to keep the number of node layers to four or fewer, so as not to unduly swamp out desired influences of input environmental parameters.

Lee (2000) similarly advocated networks with few layers, or short lengths between 'input' and 'output' nodes. This is because in networks with long path lengths, signals may be attenuated; the shorter the path between nodes, the greater the response in one to a change in the other (Lee 2000). Similarly, Kocabas and Dragicevic (2006) noted that the number of nodes in a network should be minimized, in order to reduce computational complexity. This is because as the number of nodes increases, the joint probability distribution (i.e. the probability of every possible event as defined by the values of all the variables) grows exponentially, resulting in increased computational effort required for the calculation of probabilities.

The number of nodes in a BBN provides only an approximate indication of network structure, as individual BBNs may include a variety of different types of node (potentially including decision and utility nodes). BBNs with the same number of nodes may also differ substantially in complexity, depending on the number of directed links between nodes, an aspect that was not recorded in this survey. Taking these caveats into account, the results indicated that the complexity of BBNs varied substantially between studies, with the number of nodes ranging from 3-98, with a mean of 20 ± 18 (SD). Networks with 12 or 13 nodes were most common, each accounting for 9% of examples. Most networks (60%) included 15 nodes or fewer, suggesting that the advantages of a relatively simple model structure, as recommended by Marcot (2006), are widely recognised.

Although often not stated specifically, it is likely that in the majority of cases, the authors of the publications will have played at least some role in developing the network structures presented, often by referring to the scientific literature or available data. However, 40% of publications referred to the use of expert knowledge to construct the networks, beyond that of the authors themselves. Twelve percent of studies also referred to some form of stakeholder consultation in developing or refining the networks presented (Table 1). For example, Borsuk *et al.* (2004) first conducted a survey of the relevant scientific literature, supported by a survey of stakeholders to identify the variables of particular interest. An initial network diagram was then used to elicit further expert opinion on the pattern of linkage between variables. In order to produce the most parsimonious yet realistic model, in this example each node in the network was reviewed to determine if the variable it represented was either: (1) controllable, (2) predictable, or (3) observable at the scale of the management problem. If not, then the node was removed, enabling the network to be simplified (Borsuk *et al.* 2004). As noted by Adriaenssens *et al.* (2004), a variety of techniques are available to reduce the number of nodes through a process of 'node absorption', although few authors in the publications selected here referred explicitly to these methods.

Trigg *et al.* (2000) noted the value of a prototype network for eliciting expert knowledge, but produced the prototype after analyzing the results of stepwise multiple regression analyses between cause and effect variables. In this case, to limit the size of CPTs, it was decided to limit the number of states to three and the number of parents to two, so that the maximum number of conditional probabilities to be determined for any variable was 27. This required representing some continuous variables with long tails by just three discrete states. This discretization of continuous variables is a challenge regularly encountered in BBN construction (see next section), and how such decisions are made can be expected to have a major influence on model performance, although has rarely been evaluated. Kuikka and Varis (1997) invited each individual expert to develop their own network diagram when modeling impacts of climatic change, enabling them to assess the uncertainties associated with model structure, an approach taken by very few other researchers.

In contrast, Kocabas and Dragicevic (2006) highlighted the fact that expert knowledge may be unreliable, and that instead of relying on this method, BBNs can be constructed directly from data in a learning process (Sanguesa and Burrell 2000). Algorithms for learning network structure from data are available in commercial software programs for creating Bayesian networks, such as Hugin (Hugin Expert A/S, Aalborg, Denmark; http://www.hugin.com/) and Netica (Norsys Systems Corp., Vancouver, British Columbia; http://www.norsys.com). Very few of the studies considered here employed this approach (e.g. Kocabas and Dragicevic 2006; although even here, the data used were hypothetical), and therefore there is very little evidence to evaluate its effectiveness. However, Trigg *et al.* (2000) found that automatic procedures for learning model structures from data failed to produce a meaningful causal structure, and therefore these authors expressed reservations about the value of this approach, preferring to base model structure on data analysis supported by 'common sense knowledge'. Similar reservations were expressed by Marcot *et al.* (2006), who highlighted the risk of overfitting data using this approach, resulting in spurious or misleading links between variables. Nyberg *et al.* (2006) similarly advocated use of human expertise to develop model structures, suggesting that automated procedures tend to generate relatively 'flat' networks with few intermediate variables. An alternative method is to use some pre-existing analytical model to develop network structure, an approach adopted by few researchers to date (e.g. Dorner *et al.* 2007, Lee and Rieman 1997).

Another feature of software programs such as Hugin is the ability to create an *object-oriented network* (OOBBN). This is a network that contains *instance nodes,* which represent an instance of another network. An OOBBN can therefore be viewed as a hierarchical model, making it easier to include repeated elements (such as the individual patch networks in this case) and improving the visual clarity of complex networks (Hugin 2003). Potentially this is a useful way of structuring complex models, but very few of the selected publications used this facility; examples of those that did include Shihab (2005) and Shihab and Chalabi (2007).

In common with a number of other authors, Henriksen *et al.* (2007) recommended that stakeholders be involved in the construction process of a Bayesian network, and developed a seven step procedure to achieve this (see Fig. 2). Such an approach may clearly be merited when the objective is some form of participatory decision-making process. As noted by a number of authors (e.g. Kuikka *et al.* 1999), the graphical nature of BBNs can help elicit beliefs from stakeholders or experts, but there is a variety of opinion regarding how best to achieve this, a point returned to in the following section.

Obtaining Conditional Probabilities

It is widely recognised that obtaining values to populate the CPTs is one of the main challenges to modelling with BBNs. The values that are entered will obviously have a major influence on the performance of the model, and yet obtaining appropriate values is often difficult, because of a lack of appropriate information. The problem increases with the number of directed links associated with each node.

The investigations surveyed here used a variety of different methods to obtain CPT values. As in the case of network structures, the precise role of the publication authors was often not made entirely explicit, although again it is likely that their subjective judgements will often have contributed to the identification of appropriate values. Alternative sources of information used by the publications included in this survey included (see Table 1):

- Observational or experimental evidence or data, either available directly to the authors or extracted from the scientific literature; reported in 56% of publications
- Expert knowledge (additional to that of the authors) to provide subjective probabilities, reported in 44% of studies
- Outputs of other empirical, mechanistic or stochastic models, reported in 28% of investigations (often used in conjunction with Monte Carlo simulation to derive probability distributions)
- Stakeholder consultation, referred to in 8% of studies.

It should be noted that these sources of information are not mutually exclusive, and a number of investigations drew upon more than one of these information sources, such as in the case of Lee (2000). This ability to integrate information is widely seen as one of the advantages of the technique. Pollino *et al.* (2007) noted that most studies have parameterized CPTs using either expert knowledge or data, but rarely have both of these information sources been combined in order to parameterise an individual variable.

Table 1. Summary of investigations that have used Bayesian Belief Networks (BBN) in the context of environmental science and management.

'Domain' refers to the type of problem or issue that was addressed; 'BBN structure' refers to how the structure of the network was developed; 'CPT values' refers to how the values entered in the CPTs were derived; 'Testing' refers to how the BBN model was validated or verified (see text); 'Sensitivity' indicates whether a sensitivity analysis was carried out (Y) or not (N); 'Conclusions' briefly summarizes the conclusion drawn by the authors regarding the value of BBNs as a method for addressing the problem described. (Note: the publications indicated by an asterisk do not present genuine Bayesian networks, although they are described as such by the authors. While using an approach conceptually similar to a BBN, involving the use of probability networks, these publications employed Analytica software (Lumina Decision Systems, California, USA) to perform the analyses. This software performs probabilistic analysis using random or Latin hypercube sampling, and not Bayesian inference).

Domain	BBN structure	CPT values	Testing	Sensitivity	Conclusions	Reference
Prediction of macroinvertebrate taxa in rivers	Expert judgement	Field data	Predictions tested with independent data, and compared using confusion matrices, Kappa etc.	N	Relatively good predictive success compared to other modelling approaches, but high uncertainty associated with predictions	Adriaenssens *et al*. (2004)
Land use change	Authors	Authors supported by available data	None; preliminary model only	N	Provided an effective framework for building the model	Bacon *et al.* (2002).
Water management	Authors	Judgement of authors	Data and expert knowledge	N	Outputs considered to be consistent with field knowledge (although not rested rigorously	Batchelor and Cain (1999).
Decline in fish catch	Published information and expert knowledge	Plans described for collecting expert knowledge and data	None; preliminary model presented	N	Pilot study presenting preliminary model	Borsuk *et al.* (2002)*

Table 1. Continued

Domain	BBN structure	CPT values	Testing	Sensitivity	Conclusions	Reference
Estuary eutrophication	Expert judgement supported by stakeholder consultation and literature review	Empirical and mechanistic models, with Monte Carlo or Latin Hypercube sampling	Model predictions tested against data, but not independent data	N	Approach considered to be useful	Borsuk *et al.* (2004)*
Decline in fish populations	Expert knowledge	Empirical and mechanistic models, expert knowledge, existing data	Model predictions compared with survey data, but statistical comparison was limited	N	Model predictions generally considered to be similar to field data	Borsuk *et al.* (2006)*
Water resource planning	Stakeholder consultation, authors	Existing data; subjective judgement	None; preliminary network presented	N	Considered to be a useful tool, although example presented was only illustrative	Bromley *et al.* (2005)
Participatory river basin planning	Stakeholder consultation, authors	Stakeholder consultation	Model compared with historical data; however validation of BBN part of model was limited	N	BNs are a powerful and flexible modelling tool to quantitatively describe the social aspects of planning of a water system and to improve participation in the model-building process	Castelletti and Soncini-Sessa (2007b)
Pollution to lakes and rivers	Developed according to pre-existing mechanistic model	Mechanistic model and Monte Carlo simulation	None; prototype model	N	Prototype indicated that an efficient model could be built	Dorner *et al.* (2007)

Table 1. Continued

Domain	BBN structure	CPT values	Testing	Sensitivity	Conclusions	Reference
Aquaculture impacts on shorebirds	Author	Literature search / existing data, empirical and mechanistic models, Monte Carlo simulation	None, although an additional agent-based modelling approach was performed for comparison	N	Approach not evaluated	Gibbs (2007)
Agriculture; predicting impacts of climate change on crop production	Expert judgement	Monte Carlo simulation models used to generate data; probability learning then used to populate CPTs	Compared probability distribution produced by BBN with Monte Carlo simulation output, but no statistical analysis of results	N	Problem of completing CPTs can be overcome by using Monte Carlo simulation models	Gu *et al.* (1996)
Fisheries; analysis of fisher's commitments to sustainable fisheries goals	Expert judgement, elicited using semi-structured interviews and structured questionnaires	Expert judgement and stakeholders; semi-structured interviews and structured questionnaires	No attempt at validation	N	BBN used to include stakeholder's viewpoints within the evaluation framework of fishery management procedures.	Haapasaari *et al.* (2007)
Forest management; regeneration of aspen trees	Author; based on literature review	Author; based on literature review	No comparison with actual data or testing of predictions, however an example of how BBNs might be validated is provided using the 'Consistency Analysis' method.	N	BBNs represent a valuable alternative to rule-based forest management expert systems, by incorporating uncertainty in the knowledge and input data	Haas (1991)

Table 1. Continued

Domain	BBN structure	CPT values	Testing	Sensitivity	Conclusions	Reference
Groundwater contamination	Expert knowledge and research data supported by stakeholder consultation	Expert knowledge and research data, supported by stakeholder consultation	None described	N	Focus is primarily on the use of BBNs in relation to stakeholder consultation	Henriksen *et al*. (2007)
Modelling land use change	Learned from hypothetical data	Learned from hypothetical data	No attempt at validation	Y	A number of advantages of use of BBNs for land use change modelling were highlighted	Kocabas and Dragicevic (2006)
Climate change impacts on aquatic systems	Expert judgement	Expert judgement	No attempt at validation	Y	Useful and efficient way of collecting a large knowledge base from experts	Kuikka and Varis (1997)
Fisheries; sustainability of Baltic cod management	Authors	Monte Carlo simulations, plus subjective assessments	No attempt at validation	Y	BBN used for decision analysis, to explore the sensitivity of decisions to different prior probabilities	Kuikka *et al*. (1999)
Fisheries; assessing land-use impacts on Bull trout	Author	Monte Carlo simulation used with existing data and models, plus subjective judgement of author and expert opinion	No attempt at validation; model was a 'prototype'	N	BBN used to explore extinction risk through linkage to population viability mnodels	Lee (2000)
Population viability of fish species	Stochastic population dynamics model	Stochastic population dynamics model, with Monte Carlo simulation	Evaluated in a series of test cases, but details of performance not provided	Y	Described as a promising approach	Lee and Rieman (1997)

Table 1. Continued

Domain	BBN structure	CPT values	Testing	Sensitivity	Conclusions	Reference
Impacts of forest management on habitat availability for three mammal species	Expert knowledge	Expert knowledge and scientific literature	No test of model predictions with data; instead validated in terms of how the model structured available knowledge	Y	The model structured available knowledge in a way that was considered useful	Lehmkuhl *et al*. (2001)
Behaviour of fishing vessels	Incorporation of BBN within agent-based simulation model	Agent-based simulation model	Hypothetical example presented	N	Illustrated potential application of the approach	Little *et al*. (2004)
Land use change; use of BBNs as knowledge representation for agents in multi-agent land-use model.	Authors	A sampling procedure is presented based on rules representing land-use decisions	No attempt at validation; models presented are illustrative only	N	BBNs presented as a means of knowledge representation for agents in a multi-agent land-use system	Ma *et al*. (2004); see also Ma *et al*. (2007)
Habitat relationships of rare species	Expert knowledge	Expert knowledge and field data	Accuracy tested against field data using confusion matrices and ROC curves	Y	The example model presented predicted species presence, but incurred a high error rate in predicting species absence.	Marcot (2006)
Habitat and population viability of fish and wildlife	Published data and expert knowledge	Expert knowledge and empirical data	None presented.	Y	Found to be useful, particularly where empirical data are lacking	Marcot *et al*. (2001)

Table 1. Continued

Domain	BBN structure	CPT values	Testing	Sensitivity	Conclusions	Reference
Decision-model for conservation management decisions	Expert knowledge	Expert knowledge	Consisted of ensuring that all evaluation guidelines were fully and correctly represented, eg through peer review.	Y	Models provided an intuitive means of exploring implications of data and uncertainty in an effective adaptive management process.	Marcot *et al.* (2006)
Water resources management	Stakeholder consultation	Existing data, expert opinion	Ongoing validation process described, involving consultation with stakeholders, but results of validation not presented	N	Considered to be a useful tool	Martín de Santa Olalla *et al.* (2005), (2007)
Habitat evaluation for mammal (caribou)	Expert knowledge supported by data where available	Expert knowledge supported by data where available	Qualitative assessment by experts; field verification undertaken but not reported	Y	Considered to be useful	McNay *et al.* (2006)
Sustainable use of tropical forest	Authors	Socio-economic data	Model predictions tested against independent data set, using regression	N	Model considered successful in terms of predictions being statistically related to independent data	Newton *et al.* (2006)
Conservation management	Authors	Field data, scientific literature and expert judgement	Illustrative models presented; no validation attempted	N	Approach considered to have potential	Newton *et al.* (2007)

Table 1. Continued

Domain	BBN structure	CPT values	Testing	Sensitivity	Conclusions	Reference
Adaptive management of forests	Expert knowledge	Expert knowledge supported by existing data	Results of a field experiment were used to update the model, but results of model tests not presented	N	Suggested that BBNs can benefit most approaches to adaptive management	Nyberg *et al.* (2006).
Stormwater pollution management	Authors	Existing data	Confusion matrix (contingency table) created, to estimate classification accuracy	N	Accuracy of 80-94% achieved; approach considered to be useful	Park and Stenstrom (2006)
Fishery management	Expert knowledge	Expert knowledge and existing data (including learning from data)	Predictions tested against separate data set and error rates found to be low (< 6%); also evaluated by experts	Y	Approach enabled expert knowledge and data to be combined using a robust, iterative approach.	Pollino *et al.* (2007)
Impacts of management on habitat availability for mammal species	Expert knowledge, authors	Existing data (including GIS analysis)	None	Y	No evaluation of modelling approach presented	Raphael *et al.* (2001)
Water quality	Author	Expert judgement (author), existing data	None; model presented is illustrative only	N	The BBN approach is recommended as a predictive model to guide Neuse River decision making because uncertainty, or accuracy, is believed to be an essential attribute for a predictive model.	Reckhow (1999)

Table 1. Continued

Domain	BBN structure	CPT values	Testing	Sensitivity	Conclusions	Reference
Management impacts on (salmonid) fish populations	Authors (13)	Expert judgement, existing data, predictions of process-based models	None	Y	BBN model judged to provide insight into the potential effects of management options	Rieman *et al.* (2001)
Habitat suitability for wolverines	Model derived from previous research (Raphael *et al.* 2001)	Expert knowledge	Tested with independent observational data. Model performance tested using Akaike's information criterion.	N	BBN performed well	Rowland *et al.* (2003)
Management of agroecosystems	Not specified	Questionnaire survey of land users; learned from data	None	N	More intuitive than traditional question-based surveys	Ryan *et al.* (2007)
Integrated management of dryland salinity	Authors	Actual and simulated GIS data	None; hypothetical example	N	Illustrated potential application of the approach	Sadoddin *et al.* (2005)
Wastewater treatment problem diagnosis	Authors and mechanistic model	Mechanistic model plus Monte Carlo simulation	Evaluated with respect to mechanistic model	N	Successfully diagnosed problems in simulations	Sahely and Bagley (2001)
Groundwater quality	Expert knowledge, including authors	Monitoring data	Tested with additional data, although results not reported.	N	Presented as a preliminary study with no evaluation of the method	Shihab (2005)
Groundwater quality	Learned from data (using Hugin software)	Learned from monitoring data	Compared BBN and OOBN output using KL-divergence	Y	Results compared favourably with a neural network technique	Shihab and Chalabi (2007)

Table 1. Continued

Domain	BBN structure	CPT values	Testing	Sensitivity	Conclusions	Reference
Habitat management of an endangered mammal	Expert judgement and scientific literature	Expert judgement, field survey and GIS data	Accuracy tested using field data using error matrix method, and Kappa statistic	Y	Overall accuracy of model predictions was 89%, indicating a moderate-high level of discrimination	Smith *et al.* (2007)
GIS-based decision-making system	Authors	Analytical formula together with Monte Carlo method, and expert knowledge	Used separate data to train and test the BBN	Y	Presented as a preliminary study to illustrate the potential value of the approach	Stassopoulou *et al.* (1998)
Management of buffer zones around water bodies	Expert judgement	Expert judgement	None	Y	No conclusions regarding method	Tattari *et al.* (2003)
Management of coastal lakes	Literature review and stakeholder consultation	Data analysis; literature review; simulation using existing models; expert opinion.	Not systematically validated; subjected to review by experts and stakeholders	N	Approach enabled the rapid and relatively easy integration of complex and diverse processes. No evaluation of model performance presented.	Ticehurst *et al.* (2007)
River acidification	Data analysis and common knowledge	Derived from data	Data set divided into two, one for model parameterization and one for testing. Results compared with results from regression analyses.	N	Compared favourably with multiple regression methods	Trigg *et al.* (2000)

Table 1. Continued

Domain	BBN structure	CPT values	Testing	Sensitivity	Conclusions	Reference
Fisheries management	Expert knowledge	Expert knowledge and mechanistic model	Outputs compared with earlier estimates	N	Substantial differences recorded between model predictions and earlier estimates	Uusitalo *et al*. (2005)
Lake management	Authors	Judgement of authors, available data	No validation presented.	Y	Approach considered to be effective	Varis *et al*. (1990)
Ecosystem mapping	Authors	Authors, scientific literature	Predictions compared with reference data, using Kappa	N	Percentage correct values produced by the BBN were low (48% and 50%) but comparable to or better than belief matrix approach	Walton and Meidinger (2006).
Water contamination	Authors judgement	Authors judgement	None; preliminary model presented	Y	Approach considered to have potential, but not tested; pilot study	Dawsey *et al*. (2006)

These authors explored how different information sources can be combined and weighted, another feature offered by BBN software programs that has been relatively little used by researchers. Similarly, the creation of CPTs through automated learning from data was employed by only a minority of studies (10%).

BBNs are generally considered to be particularly useful when expert knowledge is an important part of the knowledge available for the chosen problem (Uusitalo *et al.* 2005), and they offer powerful tools for assisting in the elicitation and analysis of expert knowledge. The problems of using expert judgement are widely acknowledged, giving rise to potential biases and error (Morgan and Henrion 1990). Such errors may arise because individual 'experts' may have an imperfect knowledge of a domain, they may lack the skills to assess probabilities accurately, or may be biased in their beliefs. However, the use of expert knowledge may not only be unavoidable when exploring the management of natural resources, it may even be considered desirable, given the complexity of the issues involved (Lee 2000). Despite the serious flaws that may be associated with subjective judgements, there may be no alternative source of information available (Morgan and Henrion 1990). As noted by Reckhow (1999), established techniques exist for eliciting expert knowledge (Morgan and Henrion 1990, Meyer and Booker 1991), including techniques for eliciting probabilities (see von Winterfeldt and Edwards 1986, Cooke 1991, Meyer and Booker 1991), but these have not been widely employed in BBN investigations to date. Reckhow (1999) also noted that there is evidence that people are generally not particularly good at representing their knowledge in the form of probabilities, indicating the need for using appropriate and effective elicitation techniques is order to achieve a successful outcome.

Pollino *et al.* (2007) noted that elicitation of parameters can be a difficult and time-consuming task, and often the knowledge of experts is incomplete. Sahely and Bagley (2001) reported that experts may provide widely ranging estimates for particular relationships, making the identification of a single probability estimate problematic. As illustration of this, Uusitalo *et al.* (2005) found that a range of different probability distributions were produced by different experts, and explored this through reference to the following questions listed by Morgan and Henrion (1990):

- Are there different disciplinary perspectives involved?
- Do different experts interpret the world with different theoretical models?
- Are there disagreements about the validity of various experiments or data sets?
- Have some experts ignored evidence that other experts consider very important?
- Are motivational biases operating?
- Are some (or all) of the experts just not very 'expert'?
- Are the questions posed simply impossible for human experts to answer?

The data available for a particular domain will often be partial, and not presented in the form of probabilities. In such cases, some degree of interpretation and subjective judgement will be necessary (eg see Haas 1991). For example, it will usually be necessary to discretize variables, by defining a number of discrete states for each variable (this reflects the computational difficulties of performing Bayesian inference with continuous variables). An important question is how sensitive is the model performance to errors in this discretization, an issue rarely examined in the publications reviewed here. Pollino *et al.* (2007) provide a useful description of how they achieved this in practice, by using recognised classifications, management thresholds or guidelines where possible, or by using expert knowledge when these were not available. The number of 'states' or 'classes' assigned to each variable were evaluated and assigned on an individual basis.

Lee and Rieman (1997) provide an example where the BBN was created as a version of a pre-existing model (in this case a mechanistic model of population dynamics). The relationships between variables were represented in the BBN by the use of Monte Carlo simulation to generate probability distributions. In such cases the BBN can be considered as an alternative formulation of an existing model, enabling the quantitative model to be integrated with qualitative expert judgement. Dorner *et al.* (2007) describe a similar approach, in this case for modelling pollution impacts on water bodies. Sahely and Bagley (2001) and Gu *et al.* (1996) provide examples of using a mechanistic model to generate CPTs, in conjunction with Monte Carlo simulation techniques. As noted by Sahely and Bagley (2001), using a model to generate conditional probabilities enables values to be calculated rapidly, a full range of possibilities to be examined, and the states and ranges of the variables to be changed during network construction. However, the conditional probability values will only be as good as the model, and therefore Sahely and Bagley (2001) suggest that only models that have been experimentally validated for the system of interest should be used.

Model Testing

A key question relating to any environmental model is 'how well does it work?' (Newton 2007). It is helpful to differentiate two kinds of model testing procedure (Shugart and West 1980): *verification*, in which a model is tested to determine whether it can be made consistent with some set of observations, and *validation*, in which a model is tested for its agreement with a set of observations that are independent of those observations used to structure the model and to estimate its parameters. It is important that the data used to test a model in this way are genuinely independent from the data used to parameterize it.

Marcot *et al.* (2006) emphasize the importance of testing and validation of BBN models to ensure that model structures and both prior and conditional probabilities are correct. In practice, many of the BBNs presented in the literature were not tested in any way, and very few were tested against genuinely independent data (Table 1). This partly reflects the fact that many publications presented networks that were purely illustrative, rather than operational. It is recognized that validation of environmental models can often be difficult, because of the lack of long-term data describing the processes being investigated (Newton 2007). Statistical procedures typically used to test environmental models include the kappa statistic, contingency tables (or the 'confusion' matrix; see Marcot *et al.* 2006) and receiver–operator characteristic (ROC) curves (Gardner and Urban 2003, Marcot *et al.* 2006). While these methods have all been used to test BBN models, their use has been rare to date (Table 1).

BBN testing has typically been performed by evaluating the predictive accuracy of the BBN model (in those cases where the model is designed to be predictive; this is not always the case). This may be achieved by measuring the frequency with which the predicted node state (that with the highest probability) is observed, relative to the actual value (Pollino *et al.* 2007). A number of authors have recognized the importance of dividing data into two groups, the first used for developing the model (eg for model calibration or training) and the second for testing (although this is not always achievable in practice; e.g. Borsuk *et al.* 2004). Examples of approaches adopted include:

- Pollino *et al.* (2007) randomly split the available data such that 80% were used for model training and 20% used for testing.
- Adriaenssens *et al.* (2004) adopted a similar approach, with respective values of 90% and 10%.
- Trigg *et al.* (2000) randomly divided data into two equal subsets, and the model developed on each subset was tested against the other subset of data. Each model was also tested on the subset of data used for its development. Results of these different tests were analysed by correlation.
- Rowland *et al.* (2003) tested the models developed by using an independent dataset to evaluate model performance. Model predictions were compared with field data using standard statistical tests (such as correlation).

Borsuk *et al.* (2004) noted that most of the goodness-of-fit statistics that are widely used for model testing relate to single-valued predictions. These authors suggested that probabilistic outputs of a Bayesian network require different methods for evaluation, such as those developed for assessing probabilistic weather predictions, which characterise different attributes of the joint distribution of predictions and observations (Murphy and Winkler, 1987). However, these have not been widely used to date for testing BBN models. A similar observation was made by Marcot *et al.* (2006), who recommended that error rates and confusion-matrix outcomes should be based on comparing actual values of predicted probabilities instead of just the most probable outcome states, as otherwise false-negative outcomes may be overstated.

The vast majority of authors considered that the BBN models they developed were in some way useful (Table 1), bringing to mind the familiar modelling axiom that 'all models are wrong, but some models are useful' (G. E. P. Box, cited in Ryan 1997). This raises the question, though, of how such 'usefulness' can be objectively evaluated, particularly if rigorous testing of the model has not been carried out. Uusitalo *et al.* (2005) provide one of the very few examples of BBN model predictions being different from comparison data, which was attributed to a change in expert opinion compared to previous estimates. Lehmkuhl *et al.* (2001) recognized the difficulty of validating (in their case) BBN-based habitat models because of the paucity of data and the risks of tautology if expert knowledge is used. They therefore suggested that model validity be assessed using other criteria, such as how the model structured available knowledge.

One potentially valuable approach, reported by very few of the studies presented here, would involve comparison of BBNs with other modelling approaches. For example, Gibbs (2007) compared a BBN with an agent-based complex systems model for examining the impacts of aquaculture on shorebirds. The value of such a comparative approach is further illustrated by the investigation of Trigg *et al.* (2000), who found that multiple regression methods predicted variables better than a BBN when tested on dependent data, but the converse was found when tested on independent data. This reflects the fact that the regression models fitted to outlying data, which were ignored in the BBN. In a series of additional tests, BBNs outperformed multiple linear regression, because of the ability of the BBN to represent non-linear relationships (Trigg *et al.* 2000). However performance of BBN models was limited by the amount of data available, which restricted the number of probabilistic links that could be made between variables, and the number of possible discrete states that each variable could assume. As a result, Trigg *et al.* (2000) concluded that the potential of the BBN approach can only be fully realized if sufficiently large data sets are available for the reliable estimation of CPTs.

Another approach was presented by Rowland *et al.* (2003), who used Akaike's information criterion to identify the most parsimonious models (those with optimal balance of fewest parameters and good fit to the data) among a competing set of models. As noted by these authors, values of Akaike's information criterion provide an indication of the goodness of fit of a model to the data by incorporating the log-likelihood, and also reflect the parsimony of the model by incorporating the number of parameters.

Sensitivity Analysis

It is widely recognized that in whichever approach to environmental modelling is used, there is value in conducting a sensitivity analysis, in which the model is run with a range of parameter estimates, and the effect of changing these values on the model outputs is observed to evaluate how robust the conclusions are (Newton 2007). Often there will be a degree of uncertainty about many of the values included in the model. The potential influence of this uncertainty can be explored through sensitivity analysis, typically by varying each of the uncertain parameters in turn, recording the response of the model, while holding all other parameters constant at their most likely values. Some 36% of studies surveyed here reported some form of sensitivity analysis (Table 1) performed with the BBNs developed.

In the context of a BBN, sensitivity analysis is used to measure the sensitivity of changes in probabilities of nodes when parameters and inputs are changed (Pollino *et al.* 2007). As noted by Kocabas and Dragicevic (2006), sensitivity analysis is useful to identify which parts of the model are critical and which are less likely to be important to the results. In their study, Kocabas and Dragicevic (2006) examined sensitivity by incorporating different numbers of nodes into the model, enabling them to evaluate what they refer to as 'node sensitivity'. This is an aspect considered by few other authors, despite the fact that model output can be highly sensitive to the number of nodes included (Kocabas and Dragicevic 2006).

Typically, most authors performed sensitivity analyses by changing values in different parts of the model (including the values in the CPTs) and assessing the effects on model output and performance (e.g. Kuikka and Varis 1997, Kuikka *et al.* 1999). For example, Varis *et al.* (1990) explored the sensitivity of the model outcome to the input variables by inputting different probability values to the input variables. Pollino *et al.* (2007) performed two types of sensitivity analyses when evaluating the network that they developed: (i) ''sensitivity to findings'', which considers how the BBN's posterior distributions change under different conditions, and (ii) ''sensitivity to parameters'', which considers how the BBN's posterior distributions change when parameters are altered. Pollino *et al.* (2007) noted that researchers tend to employ only one or the other of these methods, rather than both.

Marcot (2006) described the use of sensitivity analysis at each stage in the model-building process to evaluate model structure and performance, and noted that such analyses can be conducted on parts of a BBN as well as on the overall model. This author also noted that sensitivity analysis can help experts evaluate whether the model is performing according to their beliefs, and can be used to help guide the collection of field data for model validation. On the basis of sensitivity analyses that they performed, Lee and Rieman (1997) found that the BBN exhibited less sensitivity to changes in parameters than alternative simulation approaches such as analytical models. This may be a general tendency of BBN models, because of the way that a signal (i.e. the change in an output due to a change in input) is attenuated as it passes through a probabilistic network, as parameter nodes are connected to intermediate nodes through probabilistic relationships (Lee and Rieman 1997).

Some software programs now provide tools to assist in performing sensitivity analyses. For example Netica (Norsys) calculates sensitivity by calculating the degree of entropy reduction (Marcot *et al.* 2006). Lehmkuhl *et al.* (2001) provide an example of using this approach. Bednarski *et al.* (2004) provide further guidance on performing a sensitivity analysis of a BBN.

Use of BBNs to Support Decision-Making

McCann et al. (2006) highlighted the fact that BBNs offer a uniquely valuable tool for supporting decision-making, by being able to instantly recalculate and display probabilities of conditions and outcomes as alternative decisions are specified. BBNs enable the risk associated with uncertainty, variability and complexity associated with potential management activities to be identified and explored, for example by comparing the probability of different outcomes arising from alternative management decisions (McCann *et al.* 2006).

While many of the BBNs described in the literature were designed to address some environmental management problem, relatively few incorporated decision or utility nodes. Examples included Haapasaari *et al.* (2007), Kuikka *et al.* (1999), Ma *et al.* (2004), Gu (1996) and Nyberg *et al.* (2006). Marcot *et al.* (2006) considered the use of BBNs as decision-support tools in some depth, highlighting the fact that they are relatively intuitive, easy to operate, and clearly display the links between data and potential decision outcomes. In this case, BBNs were used to consolidate information and to help support discussion, rather than provide predictive models. These authors also noted that the fact that BBNs show outcomes as probabilities fits well within a risk management framework, and is often appreciated by decision-makers. Kuikka and Varis (1997) similarly found that a key asset of BBNs to decision makers is that the method provides a sense of the relative likelihood of different outcomes.

Nyberg *et al.* (2006) provide a detailed consideration of the use of BBNs to support adaptive management, highlighting the value of BBNs for:

- (i) graphically representing the structure of the domain being managed, including linkages between management actions, system components and outcomes;
- (ii) exploring the effects of alternative management actions on outcomes, by forecasting responses of variables (or indicators);
- (iii) identifying key uncertainties or gaps in understanding;
- (iv) assessing the sensitivity of forecasted outcomes to different inputs, actions or beliefs;

(v) providing a means of documenting current understanding of the domain and communicating this to stakeholders;
(vi) selecting management actions to be compared using experimental approaches;
(vii) identification of suitable indicators to provide a basis for monitoring;
(viii) comparing monitoring results with forecast system responses;
(ix) refining the model using monitoring data, by updating conditional probabilities.

The value of BBNs as decision support tools is evidenced by their widespread use in other sectors, such as medicine and commerce (Fenton and Neil 2007). In the field of environmental management, relatively few BBNs appear to have been developed as operational decision support systems, Newton *et al.* (2006) providing one example. This may reflect the fact that the literature search focused on the results of research investigations, rather than seeking evidence of BBNs being used operationally. Alternatively, it may be the case that the value of BBNs for developing decision-support systems is not yet widely appreciated in the environmental sector.

Advantages of BBNs

The value of BBNs as a modelling approach will clearly depend on the use to which they are put. Castelletti and Soncini-Sessa (2007a) suggest that there are three main uses of BBNs: (1) for modelling, when they are used to describe and explore the domain being studied; (2) to support decision making, when they may include decision and utility nodes, and might be employed as a decision support system (see previous section), and (3), as a visualization tool to summarize simply the outcomes of more complex models. Most of the examples cited here focus on the first two uses, but a number of studies addressed the third.

Any modelling approach has both advantages and disadvantages. Among the positive features of the BBN decision models that they developed, Marcot *et al.* (2006) considered that had they:

- helped organized thinking;
- allowed organization of complex relationships, and helped prompt new insights;
- provided rapid testing of the influence and value of new knowledge, and the effects of uncertainty;
- identified where there was limited or insufficient information and uncertainty; and
- provided a rigorous basis for eliciting expert opinion.

One of the key features of BBNs that is recognized by a variety of authors is their transparency, which makes the assumptions made in constructing the model readily apparent (Lee and Rieman 1997). This transparency is supported by their graphical nature, which helps provide a focus for discussion among researchers, experts, stakeholders and decision-makers (Bromley *et al.* 2005). The technique therefore encourages interdisciplinary discussion and stakeholder participation, a quality that has been widely appreciated (Batchelor and Cain 1999). Another key feature is the ability to represent uncertainty, an attribute that is widely appreciated in the environmental sector, where information is often lacking and the relationships between different variables is often poorly understood. BBNs can help to provide an understanding of the sources and implications of uncertainties that exist in both data and expert knowledge (Smith *et al.* 2007). BBNs are also considered to be particularly useful for communicating risk and uncertainty to decision makers (Sadoddin *et al.* 2005). BBNs can integrate quantititative data, including observations and results from model simulations, together with qualitative information. This attribute has attracted the interests of researchers interested in integrating socio-economic with biophysical information, as is often required when investigating environmental management. As noted by Batchelor and Cain (1999), while belief networks are no substitute for high quality field data, they provide a mathematical framework that facilitates interdisciplinary data capture and analysis.

The ability to readily incorporate expert knowledge is a further positive attribute, particularly in domains where quantitative information is lacking (Sadoddin *et al.* 2005). Smith *et al.* (2007) highlighted the fact that environmental managers may not have access to robust scientific data on which to develop their management plans and actions; in such circumstances, they are likely to be highly dependent on expert knowledge and opinion (Cain 2001, Martin *et al.* 2005). Expert knowledge can therefore be considered as a valuable resource for ecological modelling (Cain 2001), as demonstrated by a number of the examples considered here (Marcot *et al.* 2001, 2006; Raphael *et al.* 2001; Taylor 2003; Clark 2005; Martin *et al.* 2005; Marcot 2006). The value of BBNs for eliciting and integrating expert knowledge is widely acknowledged in such studies (Smith *et al.* 2007).

The use of BBNs as a tool to support stakeholder consultation and analysis is illustrated by the work of Henriksen *et al.* (2007), who concluded that the method offers a transparent, inclusive, coherent and equitable methodology for exploring environmental management, in a way that incorporates the institutional arrangements of different stakeholder groups. In particular, Bayesian networks enable the different values, interests and beliefs of various stakeholders to be explored in a participatory process in areas where existing analytical models are inadequate, because of lack of data or knowledge (Henriksen *et al.* 2007). Similarly, Bromley *et al.* (2005) highlighted the value of Bayesian networks for stakeholder consultation and as a decision support tool, highlighting the fact that the impact of a number of potential actions, or combination of actions, can be simulated very quickly. In addition, the use of networks encourages a holistic approach to management planning, as all relevant aspects of a problem can be represented in the model (Bromley *et al.* 2005).

Other advantages listed by different authors included:

- McCann et al. (2006) highlighted the value of BBNs for representing the complexity of ecological and resource management systems, enabling the problem to be partitioned and structured, and enabling value-laden concepts to be clearly represented by empirical parameters.
- Gu *et al.* (1996) noted that BBNs can be used to answer 'backwards' questions, such as if a high yield of a crop is to be achieved in a future year, what kind of management decisions would need to be made now?
- Smith *et al.* (2007) suggested that BBNs provide a relatively low cost modelling method, in that expensive computer programming or modelling expertise are not required to develop and update models.

- Dorner *et al.* (2007) considered that BBNs are very rapid in terms of running simulations.
- Marcot (2006) cited the value of being able to use case data to update model probabilities or model structure, using automated learning methods
- Sahely and Bagley (2001) noted that a major advantage is that variables may be added or the number of states for a variable changed without modifying the entire network
- Sadoddin *et al.* (2005) noted that any change in the likelihood of a state of a variable in the BN is propagated through the network. The state of the entire domain can be estimated given changes in any part of the model, in a way similar to neural networks.

Limitations of BBNs

A number of authors identified limitations or problems using BBNs. For example, Gu *et al.* (1996) highlighted the difficulties of estimating probabilities in domains where neither sufficient data nor sufficient human expertise are available. These authors also suggested that it can be a difficult and time-consuming task to describe a problem using a BBN, especially when the problem is relatively complicated, although this problem was not widely cited by other authors. However, the problem of defining a large number of probabilities to characterize the relationships between different variables is very widely acknowledged (see earlier section); the problem increases combinatorially with the number of parent nodes (Haas 1991). Lauritzen and Speigelhalter (1988) suggested that this problem may limit the development of very large BBNs, a contention that is borne out by the examples cited here, most of which included a relatively small number of nodes. Gu *et al.* (1996) also noted that in some environmental domains, the interactions among variables may be highly complex and difficult to quantify, even subjectively. Other widely acknowledged problems include the subjective nature of much expert knowledge, and the discretization of continuous variables, as noted earlier.

The fact that BBNs are poorly suited to examine dynamics over time was widely recognised. A potential method of overcoming this problem is the use of Dynamic Bayesian Networks (DBN) (Brandherm and Jameson 2004), as described by Shihab and Chalabi (2007), by representing different time events in an OOBBN framework. Nyberg *et al.* (2006) opined that such methods are relatively cumbersome compared to other modelling approaches. Furthermore, Borsuk *et al.* (2004) and Nyberg *et al.* (2006) highlighted the inability of BBNs to explicitly represent system feedbacks; relationships represent either influences at a particular instant in time or net influences on eventual steady-state conditions.

Problems or limitations highlighted by other authors included:

- The representation of algebraic relations in a BBN can be very onerous (Castelletti and Soncini-Sessa, 2007a).
- BBNs are prone to many of the general limitations common to other modelling approaches, such as the difficulty of quantifying or comparing societal values and

preferences; it is also difficult to incorporate all sources of uncertainty and variability in a model without errors and inaccuracies (McCann *et al.* 2006).

- Lee (2000) noted that there is a loss in precision when using BBNs compared to traditional approaches to population viability analysis (PVA), involving stochastic modelling.
- Borsuk *et al.* (2004) indicated that in common with alternative modelling approaches, BBNs are subject to uncertainty in the causal structure itself, in addition to the uncertainty represented by the probability distributions incorporated in the CPTs. The model predictions will therefore be associated with a degree of uncertainty than that suggested by the model. Suggestions for addressing uncertainty in model structure include learning from additional data, Bayesian model averaging and rigorous model testing (Borsuk *et al.* 2004).
- Henriksen *et al.* (2007) suggested that BBNs are difficult to understand for non-experts without any training, although the fact that BBNs can be visualized graphically, and model structures can be highly intuitive, runs counter to this view. Henriksen *et al.* (2007) also suggested that BBNs require expert input and the involvement of other stakeholders, although this will obviously be highly dependent on the domain being modelled. In fact, useful BBN models can be developed by small teams of researchers, without external input, as indicated by some of the examples presented here (e.g. Newton *et al.* 2007).

Conclusions and Forward Look

The consensus among the 50 publications included in this review is that BBNs offer a useful approach to environmental modelling, particularly in domains where decisions need to be made under conditions characterized by uncertainty, and where modelling requires incorporation of expert knowledge. Given this, it is no coincidence that BBN approaches have proved to be of particular interest to researchers investigating management of natural resources such as fisheries, water, forests and agroecosystems. While some very substantive research projects have been completed, the literature relating to BBNs is still characterized by a relatively high proportion of illustrative or pilot models, with few studies having conducted intensive model testing against independent data sets. The positive views widely expressed about BBNs may therefore reflect a general lack of rigorous, critical evaluation of the method; it is striking how few studies have so far provided comparative analyses with alternative modelling approaches. For example, it would be interesting to compare Bayesian networks with other methods of probabilistic analysis employing similar graphical approaches, such as those employed by Borsuk *et al.* (2002, 2004, 2006).

A further limitation of the research undertaken to date is that relatively few studies have explored the full functionality of BBNs, including the capacity for automated learning of both network structures and CPT values, use and weighting of different cases, 'backwards' analysis, OOBBN, analysis of d-separation, etc. Yet the number of publications produced featuring BBNs has increased steadily over recent years (Figure 5), indicating a continuing growth in interest, and it seems likely that the functionality of BBNs will be increasingly explored as this trend continues.

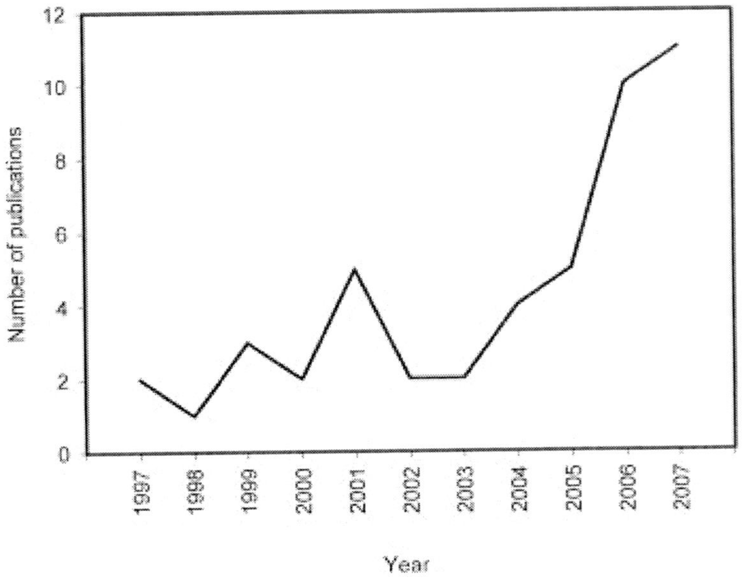

Figure 5. The number of publications produced each year, during the past decade, focusing on Bayesian Belief Networks, as identified in the literature search (see text).

It is notable that certain types of domain, namely fisheries management and water resource management, account for the majority of BBN publications produced to date. The results achieved in these domains imply that there is scope for greater use of BBNs in other areas, and the application of the method to an increasingly wide variety of environmental problems is another likely future trend. It is notable how few studies have so far been undertaken on topical issues such as climate change, invasive species, harvesting of animal populations and land cover change, to select just a few areas with potential.

Two other areas with particular potential also merit attention. The first is geoinformatics, a term that embraces techniques such as remote sensing, GIS and spatial modelling. It is surprising how few studies have explored links between BBNs and spatial data (e.g. Park and Stenstrom 2006, Kocabas and Dragicevic 2006, Stassopoulou *et al.* 1998, Smith *et al.* 2007, Walton and Meidinger 2006). Given the rapid technological developments in this area, and the rapidly increasing availability of spatial data, this is surely an area ripe for further exploration, including aspects such as image classification and scenario-building for landscape planning and management. The fact that remote sensing datasets are typically very large may help overcome the commonly cited problem of insufficient data to derive appropriate CPT values.

The second area with high potential is integration of BBNs with other modelling approaches. Some of the most successful examples of using BBNs in the recent literature involve integration with stochastic, mechanistic or process-based models, which can be used to generate network structures and to populate CPTs. In addition BBNs can be used to integrate or further explore the outputs from such models, enabling uncertainty to be explicitly represented. Examples include the integration of BBNs in a distributed modeling architecture with a multi-agent-based behavioral economic landscape (MABEL) model (Lei *et al.* 2005), the integration of a BBN with a pre-existing mechanistic model for examining non-point source pollution in watersheds (Dorner *et al.* 2007), the incorporation of a BBN within an agent-based model of fishing dynamics (Little *et al.* 2004), and use of a BBN to explore the policy implications of a stochastic population viability model (Steventon *et al.* 2006).

It is clear that BBNs will not comprise the most appropriate modelling tool in all circumstances (Bromley *et al.* 2005). For example, BBNs are not well suited for representing systems where the number of values that each variable can assume is very high, or where well established models are already available (Castelletti and Soncini-Sessa 2007a). The issue then is to define under which circumstances BBNs are likely to offer advantages over alternative techniques. This requires rigorous comparative studies to be undertaken, few of which have been reported to date. Identification of synergy between modelling approaches would also be of value, which may potentially lead to increasing integration of BBNs with other modelling methods, as noted above. Reckhow (1999) noted that any process model can be easily incorporated into a BBN model if the accuracy of the mathematical process description can be quantified and is acceptable. In a situation where such integration is achieved, the added value provided by the BBN will include the provision of a tool to foster communication between scientists, stakeholders, and decision makers; and by providing a tool to explore the uncertainty in model predictions (Reckhow 1999). If BBNs are to fulfill their undoubted potential for such applications, then as noted by Marcot *et al.* (2006), greater emphasis needs to be given in future to rigorous model testing and validation.

Acknowledgements

This research was undertaken as part of the ReForLan project, funded by the European Community's Sixth Framework Programme (FP6), contract number 032132. Thanks to Natalia Tejedor Garavito for assistance.

References

Adriaenssens, V., Goethals, P.L.M., Charles, J., De Pauw, N. (2004). Application of Bayesian Belief Networks for the prediction of macroinvertebrate taxa in rivers. *Annales de Limnologie – International Journal of Limnology,* **40(3)**, 181-191.

Bacon, P.J., Cain, J.D., Howard, D.C. (2002). Belief network models of land manager decisions and land use change. *Journal of Environmental Management,* **65(1)**, 1-23.

Batchelor, C., Cain, J. (1999). Application of belief networks to water management studies. *Agricultural Water Management,* **4(1)**, 51-57.

Bednarski, M., Cholewa, W., Frid, W. (2004). Identification of sensitivities in Bayesian networks. *Engineering Applications of Artificial Intelligence,* **17(4)**, 327-335.

Borsuk, M.E., Burkhardt-Holm, P., Reichert, P. (2002). A Bayesian network for investigating the decline in fish catch in Switzerland. In: *Integrated Assessment and Decision Support.* Proceedings of the 1st Biennial meeting of the International Environmental Modelling and Software_Society. Rizzoli, J.A.J., Lugano, A.E., editors. *Environmental Modelling and Software Society*, Manno, Switzerland, pp. 108-113.

Borsuk, M.E., Reichert, P., Peter, A., Schager, E., Burkhardt-Holm, P. (2006). Assessing the decline of brown trout (*Salmo trutta*) in Swiss rivers using a Bayesian probability network. *Ecological Modelling*, **192(1-2)**, 224-244.

Borsuk, M.E., Stow, C.A., Reckhow, K.H. (2004). A Bayesian network of eutrophication models for synthesis, prediction, and uncertainty analysis. *Ecological Modelling,* **173(2-3)**, 219-239.

Brandherm, B., Jameson, A. (2004). An extension of the differential approach for Bayesian network inference to dynamic Bayesian networks. *International Journal of Intelligent Systems,* **19(8)**, 727-748.

Bromley, J., Jackson, N.A., Clymer, O.J., Giacomello, A.M., Jensen, F.V. (2005). The use of Hugin to develop Bayesian networks as an aid to integrated water resource planning. *Environmental Modelling and Software*, **20**, 231-242.

Cain, J. (2001). *Planning improvements in natural resources management. Guidelines for using bayesian networks to support the planning and management of development programmes in the water sector and beyond.* Centre for Ecology and Hydrology, Wallingford, UK.

Castelletti, A., Soncini-Sessa, R. (2007a). Bayesian Networks and participatory modelling in water resource management. *Environmental Modelling and Software,* **22(8)**, 1075-1088.

Castelletti, A., Soncini-Sessa, R. (2007b). Coupling real-time control and socio-economic issues in participatory river basin planning. *Environmental Modelling and Software,* **22(8)**, 1114-1128.

Castillo, E., Gutierrez, J.M., Hadi, A.S. (1997). *Expert systems and probabilistic network models*. Springer, New York.

Clark, J.S. (2005). Why environmental scientists are becoming Bayesians. *Ecology Letters*, **8**, 2-14.

Cooke, R.M. (1991). *Experts in uncertainty, opinion and subjective probability in science.* Oxford University Press, New York.

Dawsey, W.J, Minsker, B.S., VanBlaricum, V.L. (2006). Bayesian Belief Networks to integrate monitoring evidence of water distribution system contamination. *Journal of Water Resource Planning and Management,* **132(4)**, 234-241.

Dorner, S., Shi, J., Swayne, D. (2007). Multi-objective modelling and decision support using a Bayesian network approximation to a non-point source pollution model. *Environmental Modelling and Software,* **22**, 211-222.

Ellison, A.M. (1996). An introduction to Bayesian inference for ecological research and environmental decision-making. *Ecological Applications,* **6**, 1036-1046.

Fenton, N., Neil, M. (2007). *Managing risk in the modern world. Applications of Bayesian networks*. London Mathematical Society, London, UK.

Gardner, R.H., Urban, D.L. (2003). Model validation and testing: past lessons, present concerns, future prospects. In *Models in Ecosystem Science*, Canham, C.D., Cole, J.J., Lauenroth, W.K., editors. Princeton University Press, Princeton, New Jersey. pp. 185-203.

Gibbs, M. T. (2007). Assessing the risk of an aquaculture development on shorebirds using a Bayesian belief model. *Human and Ecological Risk Assessment,* **13(1)**, 156-179.

Grant, W.E., Swannack, T.M. (2008). *Ecological modelling. A common-sense approach to theory and practice*. Blackwell Publishing, Malden, USA and Oxford, UK.

Gu, Y., McNicol, J., Peiris, D., Crawford, J., Marshall, B., Jefferies, R. (1996). A belief network-based system for predicting future crop production. *AI Applications,* **10**(1), 13-54.

Haapasaari, P., Michielsens, C.G.J., Karjalainen, T.P.,Reinikainen, K., Kuikka, S. (2007). Management measures and fishers' commitment to sustainable exploitation: a case study of Atlantic salmon fisheries in the Baltic Sea. *ICES Journal of Marine Science*, **64(4)**, 825-833.

Haas, T. C. (1991). A Bayesian Belief Network advisory system for aspen regeneration. *Forest Science,* **37(2)**, 627-654.

Heckerman, D. (1997). Bayesian networks for data mining. *Data Mining and Knowledge Discovery,* **1**(1), 79-119.

Henriksen, H. J., Rasmussen, P., Brandt, G., von Bulow, D., Jensen, F.V. (2007). Public participation modelling using Bayesian networks in management of groundwater contamination. *Environmental Modelling and Software,* **22(8)**, 1101-1113.

Hugin (2003). *Manual to Hugin Developer 6.3.* Aalborg, Denmark: Hugin Expert A/S.

Jensen, F.V. (1996). *An introduction to Bayesian networks*. UCL press, London.

Jensen, F.V. (2001). *Bayesian networks and decision graphs*. Springer, New York.

Kocabas, V., Dragicevic, S. (2006). Coupling Bayesian Networks with GIS-based cellular automata for modeling land use change. In: *GI Science 2006, Lecture Notes in Computer Science 4197.* Raubal, M., Miller H.J., Frank, A.U., Goodchild, M.F., editors. Springer Berlin / Heidelberg, pp. 217-233.

Kuikka, S., Hilden, M., Gislason, H., Hansson, S., Sparholt, H., Varis, O. (1999). Modeling environmentally driven uncertainties in Baltic cod (*Gadus morhua*) management by Bayesian influence diagrams. *Canadian Journal of Fisheries and Aquatic Sciences*, **56**, 629-641.

Kuikka, S., Varis, O. (1997). Uncertainties of climate change impacts in Finnish watersheds: a Bayesian network analysis of expert knowledge. *Boreal Environment Research,* **2**, 109-128.

Lauritzen, S.L., Spiegelhalter, D.J. (1988). Local computations with probabilities on graphical structures and their application to expert systems. *Journal of the Royal Statistical Society B,* **50**, 157-224

Lee, D.C. (2000). Assessing land-use impacts on bull trout using Bayesian belief networks. In: *Quantitative methods for conservation biology*. Ferson, S., editor. Springer, New York, pp. 127-147.

Lee, D.C., Rieman, B.E. (1997). Population viability assessment of salmonids by using probabilistic networks. *North American Journal of Fisheries Management,* **17**, 1144-1157.

Lehmkuhl, J.F., Kie, J.G., Bender, L.C., Servheen, G., Nyberg, H. (2001). Evaluating the effects of ecosystem management alternatives on elk, mule deer, and white-tailed deer in the interior Columbia River basin, USA. *Forest Ecology and Management,* **153(1-3)**, 89-104.

Lei, Z., Pijanowski, B.C., Alexandridis, K.T., Olson, J. (2005). Distributed modeling architecture of a multi-agent-based behavioral economic landscape (MABEL) model. *Simulation-Transactions of the Society for Modeling and Simulation International,* **81(7)**, 503-515.

Little, L. R., Kuikka, S., Punt, A.E., Pantus, F., Davies, C.R., Mapstone, B.D. (2004). Information flow among fishing vessels modelled using a Bayesian network. *Environmental Modelling and Software,* **19**, 27-34.

Ma, L., Arentze, T., Borgers, A., Timmermans, H. (2004). Using Bayesian Decision Networks for knowledge representation under conditions of uncertainty in multi-agent land use simulation models. In: *Recent Advances in Design and Decision Support Systems in Architecture and Urban Planning*. van Leeuwen, J.P., Timmermans, H.J.P., editors, Kluwer, Netherlands. pp.129-144.

Ma, L., Arentze, T., Borgers, A., Timmermans, H. (2007). Modelling land-use decisions under conditions of uncertainty. *Computers Environment and Urban Systems,* **31(4)**, 461-476.

Marcot, B.G. (2006). Characterizing species at risk I: Modeling rare species under the Northwest Forest Plan. *Ecology and Society,* **11**(2), Online URL www.ecologyandsociety.org/vol11/iss2/art10/.

Marcot, B.G., Steventon, J.D., Sutherland, G.D., McCann, R.K. (2006). Guidelines for developing and updating Bayesian belief networks applied to ecological modeling and conservation. *Canadian Journal of Forest Research,* **36**, 3063-3074.

Marcot, B.G., Hohenlohe, P.A., Morey, S., Holmes, R., Molina, R., Turley, M.C., Huff, M.H., Laurence, J.A. (2006). Characterizing species at risk II: using Bayesian belief networks as decision support tools to determine species conservation categories under the Northwest Forest Plan. *Ecology and Society,* **11(2)**, 12. [online] URL: http://www.ecologyandsociety.org/vol11/iss2/art12/

Marcot, B.G., Holthausen, R.S., Raphael, M.G., Rowland, M.M., Wisdom, M.J. (2001). Using Bayesian belief networks to evaluate fish and wildlife population viability under land management alternatives from an environmental impact statement. *Forest Ecology and Management,* **153 (1-3)**, 29-42.

Martin, T.G., Kuhnert, P.M., Mengersen, K., Possingham, H.P. (2005). The power of expert opinion in ecological models using Bayesian methods: impact of grazing on birds. *Ecological Applications,* **15**, 266-280.

Martín de Santa Olalla, F., Dominguez, A., Artigao, A., Fabeiro, C., Ortega, J.F. (2005). Integrated water resources management of the hydrogeological unit Eastern Mancha using Bayesian Belief Networks. *Agricultural Water Management,* **77(1-3)**, 21-36.

Martín de Santa Olalla, F.M., Dominguez, A., Ortega, F., Artigao, A., Fabeiro, C. (2007). Bayesian networks in planning a large aquifer in Eastern Mancha, Spain. *Environmental Modelling and Software,* **22(8)**, 1089-1100.

McCann, R.K., Marcot, B.G., Ellis, R. (2006). Bayesian belief networks: applications in ecology and natural resource management. *Canadian Journal of Forest Research,* **36 (12),** 3053-3062.

McNay, R.S., Marcot, B.G., Brumovsky, V., Ellis, R.A. (2006). Bayesian approach to evaluating habitat for woodland caribou in north-central British Columbia. *Canadian Journal of Forest Research,* **36**, 3117-3133.

Meyer, M., Booker, J. (1991). *Eliciting and Analyzing Expert Judgment: A Practical Guide.* Academic Press, London.

Morgan, M., Henrion, M. (1990). *Uncertainty: a guide to dealing with uncertainty in quantitative risk and policy analysis*. Cambridge University Press, New York, Cambridge

Murphy, A., Winkler, R.L. (1987). A general framework for forecast verification. *Monthly Weather Review*, **115**, 1330–1338.

Newton, A. C. (2007). *Forest ecology and conservation. A handbook of techniques*. Oxford University Press, Oxford, UK.

Newton A.C., Stewart, G.B. Diaz, A., Golicher, D., Pullin A.S. (2007). Bayesian Belief Networks as a tool for evidence-based conservation management. *Journal for Nature Conservation* **15**, 144-160.

Newton, A.C., Marshall, E., Schreckenberg, K., Golicher, D., Te Velde, D.W., Edouard, F., Arancibia, E. (2006). Use of a Bayesian Belief Network to predict the impacts of commercializing non-timber forest products on livelihoods. *Ecology and Society*, **11 (2)**, 24. [online] URL: http://www.ecologyandsociety.org/vol11/iss2/art24/

Nyberg, J. B., Marcot, B.G., Sulyma, R. (2006). Using bayesian belief networks in adaptive management. *Canadian Journal of Forest Research,* **36**, 3104-3116.

Olson, R. L., Willers, J.L., Wagner, T.L. (1990). A framework for modelling uncertain reasoning in ecosystem management II. Bayesian Belief Networks. *AI Applications in Natural Resources Management,* **4**, 11-24.

Park, M.-H., Stenstrom M.K. (2006). Using satellite imagery for stormwater pollution management with Bayesian networks. *Water Research,* **40(18)**, 3429-3438.

Pearl, J. (1988). *Probabilistic Reasoning in Intelligent Systems: Networks of Plausible Inference.* Morgan Kaufmann Publishers, Inc., Mateo, California, USA.

Pearl, J. (1986). Fusion, propagation and structuring in belief networks. *Artificial Intelligence,* **29**, 241-288.

Pearl, J. (1995). Causal diagrams for Empirical research. *Biometrika,* **82(4)**, 669-688.

Pollino, C.A., Woodberry, O., Nicholson, A., Korb, K., Hart, B.T. (2007). Parameterisation and evaluation of a Bayesian network for use in an ecological risk assessment. *Environmental Modelling and Software*, **22(8)**, 1140-1152.

Raphael, M.G., Wisdom, M.J., Rowland, M.M., Holthausen, R.S., Wales, B.C., Marcot, B.G., Rich, T.D. (2001). Status and trends of habitats of terrestrial vertebrates in relation to land management in the interior Columbia river basin. *Forest Ecology and Management,* **153(1-3)**, 63-88.

Reckhow, K.H. (1999). Water quality prediction and probability network models. *Canadian Journal of Fisheries and Aquatic Sciences,* **56(7)**, 1150-1158.

Reckhow, K.H. (2003). Bayesian approaches in ecological analysis and modeling. In: *Models in ecosystem science*, Canham, C.D., Cole. J.J., Lauenroth, W.K., editors. Princeton University Press, Princeton, New Jersey. pp. 168-183.

Regan, H.M., Colyvan, M., Burgman, M.A. (2002). A taxonomy and treatment of uncertainty for ecology and conservation biology. *Ecological Applications,* **12**, 618-628.

Rieman, B., Peterson, J.T., Clayton, J., Howell, P., Thurow, R., Thompson, W., Lee, D. (2001). Evaluation of potential effects of federal land management alternatives on trends of salmonids and their habitats in the interior Columbia River basin. *Forest Ecology and Management,* **153**, 43-62.

Rowland, M.M., Wisdom, M.J., Johnson, D.H., Wales, B.C., Copeland, J.P., Edelmann, F.B. (2003). Evaluation of landscape models for wolverines in the interior northwest, United States of America. *Journal of Mammalogy,* **84(1)**, 92-105.

Ryan, J.G., McAlpine, C.A., Ludwig, J.A. (2007). GLAMS: a graphical method for capturing land and water management practices in agroecosystems. *Ecosystems,* **10**, 432-447.

Ryan, T.P. (1997). *Modern Regression Methods*. Wiley and Sons, New York.

Sadoddin, A., Letcher, R.A., Jakeman, A.J., Newham, L.T.H. (2005). A Bayesian Decision Network approach for assessing the ecological impacts of salinity management. *Mathematics and Computers in Simulation,* **69(1-2)**, 162-176.

Sahely, B.S.G.E., Bagley, D.M. (2001). Diagnosing upsets in anaerobic wastewater treatment using Bayesian belief networks. *Journal of Environmental Engineering-ASCE,*_**127**(4), 302-310.

Sanguesa, R., Burrell, P. (2000). Application of Bayesian Network learning methods to Waste Water Treatment Plants. *Applied Intelligence,* **13**, 19-40.

Shihab, K. (2005). Modeling groundwater quality with Bayesian techniques. *Intelligent Systems Design and Applications. Proceedings of the* **5***th International Conference on Intelligent Systems Design and Applications* **(8-10)**, 73 - 78.

Shihab, K., Chalabi, N.. (2007). Dynamic modeling of ground-water quality using Bayesian techniques. *Journal of the American Water Resources Association,* **43(3)**, 664–674.

Shugart, H.H., West, D.C. (1980). Forest succession models. *BioScience*, **30**, 308-313.

Smith, C.S., Howes, A.L., Price, B., McAlpine, C.A. (2007). Using a Bayesian belief network to predict suitable habitat of an endangered mammal - The Julia Creek dunnart (*Sminthopsis douglasi*). *Biological Conservation,* **139 (3-4)**, 333-347.

Spiegelhalter, D.J., Dawid, A.P., Lauritzen, S.L., Cowell, R.G. (1993). Bayesian analysis in expert systems. *Statistical Science,* **8(3)**, 219-283.

Stassopoulou, A., Petrou, M., Kittler, J. (1998). Application of a Bayesian network in a GIS based decision making system. *International Journal of Geographical Information Science,* **12(1)**, 23-45.

Steventon, J.D., Sutherland, G.D., Arcese, P. (2006). A population-viability-based risk assessment of Marbled Murrelet nesting habitat policy in British Columbia. *Canadian Journal of Forest Research,* **36**, 3075-3086.

Tattari, S., Schultz, T., Kuussaari, M. (2003). Use of belief network modelling to assess the impact of buffer zones on water protection and biodiversity. *Agriculture Ecosystems & Environment,* **96(1-3)**, 119-132.

Taylor, K.J. (2003). *Bayesian Belief Networks: a conceptual approach to assessing risk to habitat.* MSc thesis, Utah State University, Logan, Utah.

Ticehurst, J.L., Newham, L.T.H., Rissik, D., Letcher, R.A., Jakeman, A.J. (2007). A Bayesian network approach for assessing the sustainability of coastal lakes in New South Wales, Australia. *Environmental Modelling and Software*, **22**, 1129-1139.

Trigg, D.J., Walley, W.J., Ormerod, S.J. (2000). A prototype Bayesian belief network for the diagnosis of acidification in Welsh rivers. In: *Development and Application of Computer Techniques to Environmental Studies*. Brebbia, C.A., Ibarra- Berastegui, G., Zannetti, P., editors. Envirosoft 2000, Bilbao, Spain. WIT Press, pp. 163-172.

Uusitalo, L., Kuikka, S., Romakkaniemi, A. (2005). Estimation of Atlantic salmon smolt carrying capacity of rivers using expert knowledge. *ICES Journal of Marine Science,* **62(4)**, 708-722.

Varis, O. (1997). Bayesian decision analysis for environmental and resource management. *Environmental Modelling and Software,* **12(2-3)**, 177-185.

Varis, O. (1998). A belief network approach to optimization and parameter estimation: application to resource and environmental management. *Artificial Intelligence*, **101(1-2)**, 135-163.

Varis, O., Kettunen, J., Sirviö, H. (1990). Bayesian influence diagram approach to complex environmental management including observational design. *Occupational Statistics and Data Analysis,* **9**, 77-91.

Varis, O., Kuikka, S. (1999). Learning Bayesian decision analysis by doing: lessons from environmental and natural resources management. *Ecological Modelling,* **119**, 177–195.

von Winterfeldt, D., Edwards, W. (1986). *Decision analysis and behavioral research.* Cambridge University Press, Cambridge, U.K.

Walton, A., Meidinger, D. (2006). Capturing expert knowledge for ecosystem mapping using Bayesian networks. *Canadian Journal of Forest Research-Revue Canadienne de Recherche Forestiere,* **36(12)**, 3087-3103.

In: Environmental Modelling: New Research
Editor: Paul N. Findley, pp. 51-88
ISBN 978-1-60692-034-3

Chapter 2

EOF Regression Analytical Model with Applications to the Retrieval of Atmospheric Temperature and Gas Constituents Concentration from High Spectral Resolution Infrared Observations

Carmine Serio, Guido Masiello and Giuseppe Grieco
DIFA, University of Basilicata, Italy

Abstract

The study describes and demonstrates a methodology for the statistical retrieval of temperature and gas species concentration that uses the spectral radiance measured by new generation of high-resolution satellite-borne infrared sensors. These include, e.g., AIRS (Atmospheric InfraRed Sounder) on AQUA satellite and IASI (Infrared Atmospheric Sounding Interferometer) on the European Meteorological Operational satellite. These spectrometers are characterized by a wide band spectral coverage (645 to 2700 cm^{-1} or 3.7 to 15.5 μm) and a spectral sampling rate in the range 0.25 to 2 cm^{-1}. The performance of the retrieval scheme has been assessed on the basis of numerical exercises. Furthermore, examples of retrievals based on real spectra measured over sea surface are given to demonstrate the ability of the scheme to obtain accurate estimation of geophysical parameters. The problem of how many principal components to retain within the regression scheme has been addressed at a length and an original procedure is presented and discussed. Furthermore, the problem of statistical interdependency of retrieval and its vertical spatial resolution has been analyzed and a new index has been designed, which is capable to quantitatively deal with such an issue. The whole methodology has been derived for a generic signal-noise model and can therefore be used to design and implement retrieval algorithms also outside the specific area of high spectral resolution infrared observations

1. Introduction

Principal Component Analysis (PCA) and the related Singular Value Decomposition (SVD) theorem (e.g., see, among many others, [17]) have been called two of the most valuable results from applied linear algebra. PCA machinery is now available in many software packages, a black box that is widely used but, to some extent, poorly understood.

One of the main goal of this work is to disclose this black box and provide a fully analytical scheme for the *linear* statistical retrieval of atmospheric thermodynamic parameters from *high spectral resolution infrared observations*. A topic, this last one, which is now of foremost interest in view of the meteorological satellite modern infrared sensors which are flying or going to fly within a few years.

These include, e.g., AIRS (Atmospheric InfraRed Sounder) on AQUA (launched in April 2002) and IASI (Infrared Atmospheric Sounding Interferometer) on the European Meteorological Operational Satellite, METOP (launched in October 2006). These spectrometers are characterized by a wide band spectral coverage (645 to 2700 cm^{-1} or 3.7 to 15.5 μm) and a spectral sampling rate in the range 0.25 to 2 cm^{-1}.

The high spectral-resolution of new advanced infrared sensors should result in better coverage and significantly improved temperature and moisture soundings capabilities as compared with the current situation. However, the full exploitation of these new sensors demands for new and improved data processing algorithms.

The problem of developing and implement a stable statistical retrieval scheme for temperature, water vapor and ozone from high spectral resolution infrared observations has been addressed in this study. The scheme has been developed by considering a suitable Empirical Orthogonal Functions (EOF) regression model between data and parameters space.

In the wide general context of Meteorology, Empirical Orthogonal Functions were first introduced by Lorenz [22] and then by Obukhov [24]. These two authors coined the term Empirical Orthogonal Functions, while the term PCA has been mostly used in the context of Statistics, where the basic idea was introduced by Pearson in the early 1900's [25]. However, the first steps were taken back in the 1870s with the paper on Singular Value Decomposition by Beltrami [2]. Formal treatment of the PCA method is due to Hotelling [15] and Rao [26].

In the context of Theoretical Physics, the PCA method is equivalent to the discrete version of the Karhunen-Loéve transform [19, 21]. This gives a representation of a stochastic process as an infinite linear combination of orthogonal functions (the equivalent of Lorenz's EOF). The coefficients in the Karhunen-Loéve transform are random variables and the expansion basis depends on the process. In fact, the orthogonal basis functions used in this representation are determined by the covariance function of the process. The Singular Value Decomposition theorem applied to the covariance matrix provides the link between PCA and the Karhunen-Loéve transform.

Practical methods for computing the SVD were unknown until 1965 when Golub and Kahan [10] published their algorithm. In 1970, Golub and Reinsch [11] published a variant of the Golub/Kahan algorithm that is still the one most-used today. Since then, the application of PCA/EOF to practical retrieval problems have flourished in many applied research fields. In the context of Satellite Meteorology, EOF decomposition was first exploited by [29] for the estimation of temperature profile from infrared spectral radiance. The application to high spectral resolution infrared observations was considered by [16], who provided a general discussion of PCA as a tool for the analysis and de-noising of infrared spectra. A quite general EOF approach to the thermal radiative transfer equation inverse problem has been recently provided by [23].

However, the application of EOF *regression* to high spectral resolution infrared observations has been pioneered by Goldberg [9], who applied the method to develop an operational retrieval algorithm for AERI instrument, and Smith and coworkers, who implemented a suitable algorithm for the analysis of high spectral resolution infrared observations [30]. Grieco et al [12, 13] applied the EOF methodology to IMG and airplane-borne infrared sounders. IMG (Interferometric Monitoring of Greenhouse Gases) is a Fourier Transform Spectrometer, which flew on board the Japanese satellite ADEOS/1 [20].

As far as linear regression is concerned, PCA has been found particularly appropriate since it provides a powerful tool to reduce a complex data set to a lower dimension, so that alleviating the so-called problem of *curse of dimensionality*.

The present study wants to provide the reader with a more comprehensive analysis of the methodology which can be also used as a guide to apply EOF regression methods for the retrieval of generic atmospheric parameters. To this end we will show the derivation of the *exact* analytical form of the regression coefficients for a generic signal-noise model, and, therefore, we will provide a general framework, which can be used to develop specific user applications. An analytical form for the EOF regression coefficients, but for the much limited model of a noiseless signal model, has been recently derived by [3].

The problem of how assessing the vertical spatial resolution of the retrieved profile has been addressed by exploiting a new method, which has been recently proposed by Serio et al [27]. Moreover, particular care has also been put to the problem of how many principal components to retain within the regression scheme or, equivalently, to represent the data space. A suitable procedure has been developed, which exploits the Kaiser criterion [18], Cattell's scree test [5] and the so-called L-curve method [14].

In addition to retrieval examples based on simulation, the methodology will be exemplified by considering applications to NAST-I and IASI data for the retrieval of temperature, water vapor and ozone. NAST-I is the NPOESS (National Polar-orbiting Operational Environmental Satellite System) Aircraft Sounder Testbed Interferometer, e.g. [8]. The NAST-I data, which have been used in the present work, refer both to the CAMEX/3 and the EAQUATE field campaigns. CAMEX/3, the Convection and Moisture Experiment 3, is the Atlantic basin tropical cyclone field validation NASA ER-2 flight (during local nighttime, September 13-14, 1998) over Andros Island, Bahamas. EAQUATE, the European AQUA Thermodynamic Experiment [28], took place in Italy and England in September 2004; the data from the Proteus flight (during local nighttime 9-10 September 2004) over Southern Italy, have been considered.

IASI, the Infrared Atmospheric Sounding Interferometer is a key payload element of the METOP series of European meteorological polar-orbiting satellites. It has been developed by the French Space Agency (CNES) in the framework of a co-operation agreement with the European Center for the Exploitation of Meteorological Satellites (EUMETSAT). The first flight model was launched in 2006 onboard the first European meteorological polar-orbiting satellites, METOP-A. The data analyzed in this study have been recorded on the day 22 July 2007 during a series of overpasses over the tropical belt.

The present study is organized as follows. Section 2. describes the mathematical background. The EOF regression model is described and developed in section 3.. Section 4. discusses the performance of the method with retrieval examples based on simulation. Section 5. describes the application to real observations. Conclusions are drawn in section 6..

2. Mathematical Theory

PCA is extremely interdisciplinary, and each field has its own nomenclature, habits and notation. At now, the Jolliffe book [17] is probably the best attempt to unify various fields. Our approach tends to adopt the conceptual view of the Karhunen-Loéve theorem, which states how to transform a given stochastic process in a new orthogonal basis. Then, the necessary ingredient of the methodology is, basically, the covariance matrix of the process, whereas the SVD theorem and related machinery is used as a tool for decomposing the covariance matrix and, hence, yielding the orthogonal functions and related eigenvalues. In what follows we use PCA, EOF decomposition, Hotelling and Karhunen-Loéve transforms as synonyms.

Let $\mathbf{s} = (s_1, s_2, ..., s_N)^t$ a sample of a N-dimensional zero mean stochastic process with covariance matrix

$$\mathbf{C}_s = E\left(\mathbf{s}\mathbf{s}^t\right) \tag{1}$$

where E and t denote expectation and transposition, respectively. Then the Hotelling transform can be obtained (e.g., [7]) by Singular Value Decomposition of $\mathbf{C}_s$,

$$\mathbf{C}_s = \mathbf{U}\mathbf{S}_s\mathbf{V}^t \tag{2}$$

where $\mathbf{U}$ and $\mathbf{V}$ are orthogonal, unitary matrices of size N by N ($\mathbf{U}\mathbf{U}^t = \mathbf{I}$, and $\mathbf{V}\mathbf{V}^t = \mathbf{I}$ with $\mathbf{I}$ the identity matrix; also note that, because a covariance operator is symmetric and positive definite, then $\mathbf{U} = \mathbf{V}$); $\mathbf{S}_s$ is a diagonal matrix whose elements are the eigenvalues (singular values), positive definite, of the operator $\mathbf{C}_s$, that is

$$\mathbf{S}_s = \text{diag}\left(\sigma_1^2, \ldots, \sigma_N^2\right) \tag{3}$$

and $\sigma_1^2 > \ldots > \sigma_N^2$, where diag $(\sigma_1^2, \ldots, \sigma_N^2)$ indicates the diagonal matrix whose diagonal elements are the components of the vector, $(\sigma_1^2, \ldots, \sigma_N^2)$. In addition, the notation σ_j^2 for the singular values stresses the fact that they are positive definite.

For any N-dimensional vector $\mathbf{s}$, the Hotelling transform or EOF decomposition reads:

$$\mathbf{c} = \mathbf{U}^t\mathbf{s}; \quad \text{with its inverse :} \quad \mathbf{s} = \mathbf{U}\mathbf{c} \tag{4}$$

with $\mathbf{c}$ being the Principal Components (PC) vector of size N. The N columns of the matrix, $\mathbf{U}$ are vectors of size N, the same as the original vector, $\mathbf{s}$. They define the Empirical Orthogonal Functions. In other word, the matrix $\mathbf{U}$ defines an orthogonal space of dimension $N \times N$, in which we can develop any vector of size, N.

One important property of the Principal Components is their orthogonality,

$$E(\mathbf{c}\mathbf{c}^t) = \mathbf{S}_s = \text{diag}\left(\sigma_1^2, \ldots, \sigma_N^2\right) \tag{5}$$

which says that the variance of the j-th PC score or component is the singular value of order j. The total variance of the stochastic process, s is given by

$$E(\mathbf{s}^t\mathbf{s}) = \sum_{j=1}^{N} \sigma_j^2 \tag{6}$$

In addition to achieving uncorrelated components, the variances (that is the singular values) of the components also will be very different in most applications. For a vector s characterized by high redundancy, a considerable number of the variances will be so small that the corresponding components can be discarded altogether. Those components that are left, say the first r, constitute the truncated expansion, $\hat{\mathbf{s}}$ of s,

$$\mathbf{s} \simeq \hat{\mathbf{s}} = \mathbf{U}\mathbf{c}^* \tag{7}$$

with

$$c_j^* = \begin{cases} c_j & \text{if } j \leq r \\ 0 & \text{otherwise} \end{cases} \tag{8}$$

Expansion (7) uses the first r principal components alone, and therefore only the first r columns of $\mathbf{U}$, in practice, need to be considered. In other words, the PC analysis transforms a huge amount of correlated data into a set of statistically uncorrelated features or components, ordered according to decreasing information content. The truncation above, in addition to act as a noise filter, reduces the dimensionality of the vector, s, which is important in view of the regression problems we are going to discuss in next sections.

3. EOF Based Regression Algorithm

3.1. Data and Parameters Space, Training Data Set and Basic Definitions

The physical data and parameters we are dealing with are the upwelling spectral radiation at the top of the atmosphere and the thermodynamic state of the atmosphere. The upwelling spectral radiance is assumed to be observed through high spectral resolution infrared spectrometers, which typically sense earth's spectrum into the spectral interval 600 to 3000 cm^{-1} (3.3 to 16.6 μm). For a spectral sampling of 0.25 cm^{-1}, this gives thousands of data points, which can be arranged to form the observed radiance vector, $\mathbf{R} = (R_1, \ldots, R_N)^t$, where N is the size of the vector.

Examples of radiance vectors, we consider in this work, are shown in Fig.s (1a) and (1b), which illustrate upwelling spectral radiances as measured by the two spectrometers NAST-I and IASI, respectively.

Figure (1a) allows us to appreciate the very high spectral resolution of NAST-I. NAST-I is an Imaging Fourier Transform Spectrometer whose nominal sampling rate is $\Delta\sigma = 0.2411$ cm^{-1}. The spectral coverage is 645 to 2700 cm^{-1}, therefore the measured radiance spectra cover CO_2 emission within the 15-μm and 4.3-μm bands, H_2O emission across the 6.3-μm band, O_3 emission within the 9.6μm and 4.7-μm bands. These radiance measurements can be used to retrieve temperature, water vapor, ozone, and to estimate the amount of other trace and minor gases.

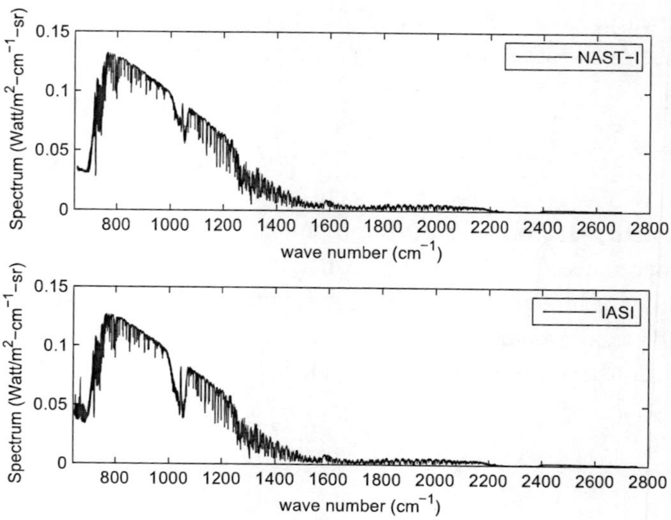

Figure 1. Example of NAST-I and IASI spectra. Both spectra have been recorded over sea surface in the tropics.

The same spectral quality is also evident in Fig. (1b), which shows an example of IASI spectrum. IASI has been designed for operational meteorological soundings with a very high level of accuracy. It has been also designed for atmospheric chemistry aiming at estimating and monitoring trace gases such as ozone, methane or carbon monoxide on a global scale. The IASI spectrometer system relies on an accurately calibrated Fourier Transform Spectrometer operating in the 3.7 - 15.5 μm spectral range. The optical configuration of the sounder is based on a Michelson interferometer.

Figure (1) exemplifies the data space we are dealing with. The kind of parameters space we are interested in is exemplified in Fig. (2), which shows a case for temperature, water vapor and ozone profile: that is what we call the state of the atmosphere.

In addition to temperature, water vapor and ozone, the state of atmosphere is specified by surface parameters (surface temperature, spectral emissivity, ground pressure) and atmospheric gas constituents profiles, that is parameters that are continuous functions of the atmospheric altitude level. We work with a discretized version of these variables, and the value of the grid step used to discretize any given profile is dictated by the accuracy of the *Forward Model,* that is the discretized version of the physical equation which governs the functional dependence of the upwelling radiance on the atmospheric state vector. In practice the discretization process involves the division of the atmosphere into M homogeneous layers and for each of these layers the mean value, within the layer, can be considered for each parameter, i.e. temperature, water vapor and so on. Then, the state vector, $\mathbf{v}$ is defined, which is made up by the individual M-dimensional vectors for temperature, water vapor, ozone and so on.

In this study, we use the so-called σ-IASI radiative transfer algorithm [1]. The forward model, σ-IASI divides the atmosphere in $M = 60$ layers.

Let y be the layer-mean value for a generic layer, and for a generic parameter of the

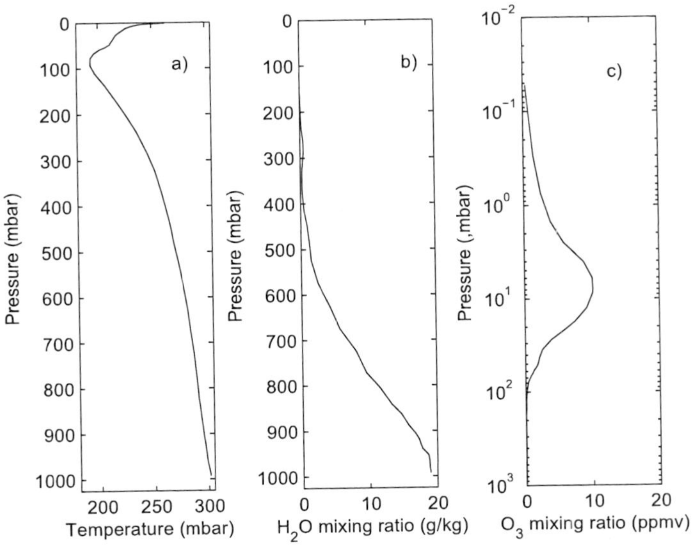

Figure 2. In contrast to the data space exemplified in Fig. (1) above, here we exemplify the parameters space. The figure shows a case of a) temperature, b) water vapor and c) ozone profile.

state vector, that is y is just one single component of the state vector, $\mathbf{v}$. We consider the linear regression problem,

$$y - \bar{y} = b_1(R_1 - \bar{R}_1) + \cdots + b_N(R_N - \bar{R}_N) \tag{9}$$

where the bar sign indicates expectation value (taken over a suitable statistical ensemble of the variate at hand) and $(b_1, \ldots, b_N)$ are regression coefficients to be determined. Note that we proceed by a layer-by-layer and a parameter-by-parameter approach, in other word we consider one single layer at a time, and for the given layer only one single parameter at a time is considered, as well. Therefore, in case we have M layers, for a given parameter, we have to consider M different regression problems. In addition, if we have a number of p parameters to be considered, we have a total of $p \cdot M$ diverse regression problems.

In practice, the observation, that is the radiance vector $\mathbf{R}$, will be affected by noise, whose variance may depend on the wavenumber, which, in turn, makes it desirable to weight each radiance element in (9) according to the standard deviation of the noise term. Denoting the noise-covariance by $\mathbf{C}_o$, we define the centered and normalized observation, $\mathbf{x}$ according to

$$\mathbf{x} = C_o^{-\frac{1}{2}}(\mathbf{R} - \bar{\mathbf{R}}) \tag{10}$$

and re-write the regression problem as,

$$y - \bar{y} = b_1 x_1 + \cdots + b_N x_N \tag{11}$$

It is quite obvious that the above regression problem is made difficult, if not unpractical, because of the large value of N.

Writing the above problem in vector notation, we have

$$y - \bar{y} = \mathbf{b}^t\mathbf{x} \tag{12}$$

Let $\mathbf{U}$ be an unitary, orthogonal transformation (so that $\mathbf{U}\mathbf{U}^t = \mathbf{I}$), then

$$y - \bar{y} = \mathbf{b}^t\mathbf{U}\mathbf{U}^t\mathbf{x} = (\mathbf{U}^t\mathbf{b})^t(\mathbf{U}^t\mathbf{x}) \tag{13}$$

which by writing, $\boldsymbol{\beta} = \mathbf{U}^t\mathbf{b}$ and $\mathbf{c} = \mathbf{U}^t\mathbf{x}$, gives

$$y - \bar{y} = \boldsymbol{\beta}^t\mathbf{c} \tag{14}$$

Provided that the transform is effective, then a few $\mathbf{c}$-components suffice to give a good representation of the vector $\mathbf{x}$, an the regression problem is reduced to

$$y - \bar{y} = \beta_1 c_1 + \cdots + \beta_r c_r \tag{15}$$

with $r << N$. In other words, to overcome the problem of high dimensionality, we consider a functional basis, decompose the variate $\mathbf{R}$ accordingly and work with a low number of coefficients. The aim is to reduce the dimension of $\mathbf{R}$ without loosing information on $y \mid \mathbf{R}$. The basis we use to decompose $\mathbf{R}$ is that of the Empirical Orthogonal Functions (EOF) derived by the covariance matrix of a suitable set or ensemble of $\mathbf{R}$. In fact, in order to be implemented, the approach above needs a suitable ensemble of pairs (hereafter *training data set*) $\{y_k, \mathbf{R}_k\}_{k=1,\dots,n}$, so that the sample covariance matrix of the covariates can be estimated and the regression coefficients computed, as well.

The training data set of state vectors consists of a suitable set of profiles for temperature, water vapor and ozone plus the surface temperature. Using a suitable forward model (and we use the σ-IASI model [1]), the synthetic spectral radiance, $\mathbf{r}$ can be computed for each state vector, $\mathbf{v}$,

$$\mathbf{r} = F(\mathbf{v}) \tag{16}$$

where F symbolically designs the forward model. Once the signal has been computed, the *observation* $\mathbf{R}$ is defined according to the additive signal-noise model

$$\mathbf{R} - \overline{\mathbf{R}} = (\mathbf{r} - \bar{\mathbf{r}}) + \mathbf{n}_e \tag{17}$$

where $\bar{\mathbf{r}}$ is the mean value of the signal, computed on the basis of the training data set, $\mathbf{R}$ is the observation and $\mathbf{n}_e$ is a zero mean Gaussian noise term. This last term models the observational noise from which any observation is in practice affected. In the context of infrared spectrometers, the assumption of Gaussianity is well established and the characterization of the noise is done through the *observational* covariance matrix, $\mathbf{C}_o$, which is assumed to be diagonal.

In contrast to what assumed for the data space, the state vector, $\mathbf{v}$, hence the parameters space, is assumed to have only intrinsic variability, that is we do not consider an additional external noise term.

As done for the observation, we introduce the centered, noise-normalized signal

$$\mathbf{s} = \mathbf{C}_o^{-\frac{1}{2}}(\mathbf{r} - \bar{\mathbf{r}}) \tag{18}$$

so that we can consider the signal-noise model in terms of standardized quantities,

$$\mathbf{x} = \mathbf{s} + \mathbf{m} \tag{19}$$

Because of the standardization, the noise term $\mathbf{m}$ has covariance matrix equal to the identity matrix, $E(\mathbf{m}\mathbf{m}^t) = \mathbf{I}$.

The noise standardization assumes an important role here, which is clearly seen when we consider the SVD decomposition of the covariance matrix of the observation, $\mathbf{x}$.

First, because of (19) we have,

$$\mathbf{C}_x = \mathbf{C}_s + \mathbf{I} \tag{20}$$

where $\mathbf{C}_s$ is the covariance matrix of the signal. Considering the SVD decomposition of $\mathbf{C}_s$ and inserting it in Eq. (20), we have

$$\mathbf{C}_x = \mathbf{U}\mathbf{S}_s\mathbf{V}^t + \mathbf{I} = \mathbf{U}\left(\mathbf{S}_s + \mathbf{I}\right)\mathbf{V}^t \tag{21}$$

which says that the EOF basis, $\mathbf{U}$ of the signal is exactly equal to that of the noisy radiance, $\mathbf{R}$, provided that we work with $\mathbf{C}_o$-normalized quantities. *This is a very important property since it says that we can compute the basis for the observations directly from the signal, without the need to effectively simulate the noisy radiance.* Another important properties is that the singular values of observations (noisy radiances) are those of the signal augmented just of one,

$$\mathbf{S}_x = \mathbf{S}_s + \mathbf{I} \tag{22}$$

therefore, if $\{\sigma_j^2, \quad j = 1, \ldots, N\}$ denotes the set of singular values for the signal, then

$$\mathbf{S}_x = \text{diag}(\sigma_1^2 + 1, \ldots, \sigma_N^2 + 1) \tag{23}$$

3.2. How Many Principal Components Do We Need to Extract?

At this point it is important to remember we are considering Principal Components Analysis as a data reduction method, that is, as a method for reducing the number of variables. The question then is, how many PC scores do we want to extract? Note that as we extract consecutive PC scores, as done in considering the truncated expansion described from the two equations (7) and (8), they account for less and less variability. The decision of when to stop extracting PC scores basically depends on when there is only very little *random* variability left.

For this problem, Eq. (23) leads us to a possible solution. In fact, it hints that an obvious choice for the cut off point, r is that r for which

$$\begin{cases} \sigma_j^2 > 1 & \text{for } j \leq r \\ \sigma_j^2 < 1 & \text{for } j \geq r+1 \end{cases} \tag{24}$$

The above criterion is quite obvious: it retains only the principal components that are above the noise level. This simply and effective idea goes back to Kaiser [18], who recommends that only eigenvalues (and, hence, associated principal components) at least equal to one have to be retained in the truncated expansion.

However, our final goal is not to best reproduce a given signal with a limited number of principal components, when rather to find a representation which is capable to minimize the *forecast error* in a problem of linear regression. As it will be shown later, this objective will bring us to a more evolute choice of r, which is much more pertinent to our problem.

3.3. The System of Regression Coefficients

If we assume that the variable y is available for n diverse realizations (e.g. from the training data set), so that we can evaluate the mean value,

$$\overline{y} = \frac{1}{n}\sum_{i=1}^{n} y_i \tag{25}$$

the regression problem (15) can be written in vector-matrix form,

$$\tilde{\mathbf{y}} = \mathbf{P}_c\boldsymbol{\beta} + \varepsilon \tag{26}$$

where ε is a zero mean, unit variance noise term and where

$$\tilde{\mathbf{y}} = (y_1 - \overline{y}, \ldots, y_n - \overline{y})^t; \boldsymbol{\beta} = (\beta_1, \ldots, \beta_n)^t \tag{27}$$

and

$$\mathbf{P}_c = \begin{pmatrix} c_{11}, \ldots, c_{1r} \\ \cdots \\ c_{n1}, \ldots, c_{nr} \end{pmatrix} \tag{28}$$

where c_{ij} is the PC score of order j ($j = 1, \ldots, r$), corresponding to the i-th observation, y_i, ($i = 1, \ldots, n$).

An estimate $\hat{\boldsymbol{\beta}}$ of $\boldsymbol{\beta}$ can be obtained by Least Squares, whose normal equations, for the case at hand, read

$$\hat{\boldsymbol{\beta}} = \left(\mathbf{P}_c^t\mathbf{P}_c\right)^{-1}\mathbf{P}_c^t\tilde{\mathbf{y}} \tag{29}$$

The above equation can be greatly simplified by taking into account the orthogonality properties of the PC transform (Eq. 5), we have

$$\hat{\boldsymbol{\beta}} = (\mathbf{S}_x)^{-1}\mathbf{P}_c^t\tilde{\mathbf{y}} \tag{30}$$

which, considering the definition of singular values for the signal, Eq. (23), can be written in component-wise form

$$\begin{aligned} \hat{\beta}_1 &= \tfrac{1}{n}\tfrac{1}{\sigma_1^2+1}\textstyle\sum_{i=1}^n c_{i1}(y_i - \overline{y}) \\ &\cdots \\ \hat{\beta}_j &= \tfrac{1}{n}\tfrac{1}{\sigma_j^2+1}\textstyle\sum_{i=1}^n c_{ij}(y_i - \overline{y}) \\ &\cdots \\ \hat{\beta}_r &= \tfrac{1}{n}\tfrac{1}{\sigma_r^2+1}\textstyle\sum_{i=1}^n c_{ir}(y_i - \overline{y}) \end{aligned} \tag{31}$$

The PC scores considered in the above equations are those of the observations, which are affected by noise, so that the problem of the statistical convergence of any of the $\hat{\beta}$'s has to be considered. However, because the noise affecting the observations is assumed to be un-correlated, so is that affecting the PC scores. Thus, the expectation value, β of any $\hat{\beta}$ can be easily computed. We have

$$\begin{aligned}
\beta_1 = E(\hat{\beta}_1) = & \ \tfrac{1}{n}\tfrac{1}{\sigma_1^2+1}\textstyle\sum_{i=1}^{n} E(c_{i1})(y_i - \overline{y}) = \tfrac{1}{n}\tfrac{1}{\sigma_1^2+1}\sum_{i=1}^{n} \tilde{c}_{i1}(y_i - \overline{y}) \\
\dots & \\
\beta_j = E(\hat{\beta}_j) = & \ \tfrac{1}{n}\tfrac{1}{\sigma_j^2+1}\textstyle\sum_{i=1}^{n} E(c_{ij})(y_i - \overline{y}) = \tfrac{1}{n}\tfrac{1}{\sigma_j^2+1}\sum_{i=1}^{n} \tilde{c}_{ij}(y_i - \overline{y}) \\
\dots & \\
\beta_r = E(\hat{\beta}_r) = & \ \tfrac{1}{n}\tfrac{1}{\sigma_r^2+1}\textstyle\sum_{i=1}^{n} E(c_{ir})(y_i - \overline{y}) = \tfrac{1}{n}\tfrac{1}{\sigma_r^2+1}\sum_{i=1}^{n} \tilde{c}_{ip}(y_i - \overline{y})
\end{aligned} \tag{32}$$

with $\tilde{c}_{ij} = E(c_{ij})$ simply given by the PC transform of the signal rather than the observation, that is

$$\tilde{\mathbf{c}} = E(\mathbf{c}) = U^t E(\mathbf{x}) = U^t \mathbf{s} \tag{33}$$

It is important here to note that the above expectations have to be considered with respect to the noise term alone. That is, we assume that the signal is kept a constant while we add external noise, so that we yield infinite copies of a given realization of the signal. These copies differ each from other only for the given noise outcome. If we considered the expectation also with respect to the training data set, then we would have the trivial result, $E(\mathbf{c}) = 0$.

To sum up, we have that the analytical form for the *true* regression coefficients depend only on the system of singular values and PC scores of the signal. Thus, in case we have a physical model for the signal, we do not need the observations for the computation of the regression coefficients, but only the second-order statistics of the signal and noise, that is their covariance matrices, $\mathbf{C}_s$ and $\mathbf{C}_o$, respectively.

Because of the noise-standardization (see Eq. 18), the effect of the noise term is just to add 1 to each singular value, σ_j^2 in Eq. (32), so that any regression coefficient has the characteristic term

$$\frac{1}{\sigma_j^2 + 1} \tag{34}$$

In case the equations (32) were derived in the limit of a noisy free signal, the above term would simply be

$$\frac{1}{\sigma_j^2} \tag{35}$$

High spectral resolution infrared radiance observations are characterized by a very high redundancy (see section 4.), and in practice the dynamic range of their singular values is incredibly wide and some of the singular values could be as small as 0, within the rounding error. In other words, while the weight (35) tends to $+\infty$ for $\sigma_j^2 \to 0$, so that amplifying any source of external perturbation to infinity, the weight (34) tends to 1 for $\sigma_j^2 \to 0$ and automatically damps the effect of exogenous noise.

3.3.1. Bias and Second Order Statistics of the Retrieval

According to our linear model (14), the true value, y_{true} of a given parameter, y, would be

$$y_{true} - \overline{y} = \boldsymbol{\beta}^t \mathbf{c}_{true} \tag{36}$$

whereas the expectation value of the estimation, $\hat{y}$, is

$$E(\hat{y}) - \bar{y} = \beta_1 c_{1,true} + \cdots + \beta_1 c_{r,true} \tag{37}$$

and, hence, the bias Δ_b is

$$\Delta_b = \mathrm{E}(\hat{y}) - y_{true} = \sum_{k=r+1}^{N} \beta_k c_{k,true} \tag{38}$$

The bias is, therefore, ultimately introduced by the truncation error in the EOF representation of the radiance vector. It is evident from (38) that the bias tends to zero as r approaches N.

To derive the second order statistics of the retrieval, consider that in our approach, the parameters are estimated once at a time. However, the predictors are the same for any given parameter, therefore if $y(i)$ is a given parameter belonging to the i-th layer, then the covariance, cov $(\hat{y}(i)\hat{y}(j))$ is expected to be different from zero. Because of the linear form of the regression model, we have

$$\text{cov}\,(\hat{y}(i)\hat{y}(j)) = \sum_{k=1}^{r} \beta_k(i)\beta_k(j)(\sigma_k^2 + 1) \tag{39}$$

where $\beta_k(i), k = 1, \ldots, r$ denotes the set of regression coefficients for the i-th layer and the generic parameter, y. Because of (39), the variance of $\hat{y}$ is

$$\text{var}(\hat{y}) = \sum_{k=1}^{r} \beta_k^2(\sigma_k^2 + 1) \tag{40}$$

which in contrast to the bias, increase with r. Thus, the mean square error,

$$\text{m.s.e.} = E\left((\hat{y} - y_{true})^2\right) = \text{var}(\hat{y}) + \Delta_b^2 \tag{41}$$

will reach a minimum. In practice, since both the singular values and the PC scores tends rapidly to zero as r becomes large, we will see a plateau rather than a sharp minimum.

The above formulas for bias and variance are of theoretical interest, since in practice the true value of the parameters is not known.

In addition, consider that the major uncertainties with the regression model at hand come from the limitation of the training data set in correctly representing the real atmosphere. This limitation can yield additional bias whose origin is here illustrated.

Let us assume that r is large enough so that there is no bias arising from the truncated PC representation of the spectral radiance, then we can write the bias as

$$\begin{aligned} \Delta_b = E(\hat{y} - y_{true}) &= E\left(\bar{y} + \beta_1 c_1 + \cdots + \beta_r c_r - y_{true}\right) \\ &= \bar{y} - E(y_{true}) + \textstyle\sum_{k=1}^{r} \beta_k E(c_k) \end{aligned} \tag{42}$$

where, now the expectation values have to be considered with respect to both the intrinsic variability (average over the training data set) and exogenous noise.

In a perfect world, where our training data set represents all the possible modes of variability of the atmosphere, then

$$\bar{y} = E(y_{true}); \quad E(c_k) = 0, \quad \text{for each}, k \tag{43}$$

and the estimate, $\hat{y}$ would be unbiased. However, in a real world, the training data set will not explain all the information, proper of a complex system such as the atmosphere. For this reason, it could be possible that the given observed radiance vector, say $\mathbf{r}_{true}$ belongs to a statistical population whose mean value, E($\mathbf{r}_{true}$) is different from that of the training data set, $\bar{\mathbf{r}}$. Consider that also in the case of a perfect training data set, we have to account for the effect of forward modeling uncertainty. Because of this uncertainty, our synthetic spectral radiance could be systematically biased with respect to the observations, which again would yield

$$E(\mathbf{r}_{true}) - \bar{\mathbf{r}} \neq 0 \tag{44}$$

Because of this difference, whatever its origin may be, the expectation values of the PC scores would be different from zero. We have

$$E(\mathbf{c}) = \mathbf{U}^t \mathbf{C}_o^{-\frac{1}{2}} \left(E(\mathbf{r}_{true}) - \bar{\mathbf{r}} \right) \tag{45}$$

To sum up, the bias can arise both from the parameters space and data space, and it is related to the limitation of our training data set to correctly reproduce the ensemble means of the real space of parameters and data. This kind of error is referred to as *generalization* error.

3.3.2. Assessing the Vertical Spatial Resolution of the Retrieval, the Index i_D

Again, the above considerations are of theoretical interests. For the purpose of retrieval performance assessment, variances and covariances can be, much more practically, computed on the basis of the training data set itself.

If the vector, $\hat{\mathbf{y}} = (\hat{y}(1), \hat{y}(2), \ldots, \hat{y}(M))^t$ is made up with the layer-mean estimated values of a given parameter, in order to form the profile function of the parameter, then the mean square error matrix can be obtained by

$$\mathbf{C}_{\hat{y}} = \mathrm{E}\left((\hat{\mathbf{y}} - \mathbf{y}_{true})(\hat{\mathbf{y}} - \mathbf{y}_{true})^t \right) \tag{46}$$

where expectation value has to be taken with respect to training data set. This matrix will be denoted with $\mathbf{C}_T$, $\mathbf{C}_{H_2O}$ and $\mathbf{C}_{O_3}$ for the case of temperature, water vapor and ozone profiles, respectively.

The retrieval covariance matrix, for a given parameter profile, can be used to analyze the spatial vertical resolution of the parameter itself. Indeed, the vector $(y(1), y(2), \ldots, y(M))^t$ represents the discretized version of a spatial function (i.e., temperature profile, water vapor mixing ratio profile, ozone mixing ratio profile). A strong correlation, that is a relatively high value of the covariance, between any two of the parameters means that the two have not been independently resolved by the data set, and that only some linear combination of the parameters is resolved. However, a direct examination and interpretation of the off-diagonal elements (covariances) of a covariance operator is not easy, which makes the definition of some suitable scalar index high desirable.

To this end, it has to be considered that the retrieval correlation is determined by the non-null off-diagonal terms in the covariance operator. For a full independent retrieval (and therefore for a retrieval, which attains the maximum possible vertical spatial resolution), the covariance matrix has to be fully diagonal.

Following Serio et al [27], we assume that the covariance operator has been normalized in order to obtain the correlation matrix,

$$C_{\hat{y}}(i,j) \leftarrow \frac{C_{\hat{y}}(i,j)}{\sqrt{C_{\hat{y}}(i,i)C_{\hat{y}}(j,j)}} \tag{47}$$

$\mathbf{C}_{\hat{y}}$ may be additively decomposed in its diagonal and off-diagonal components:

$$\mathbf{C}_{\hat{y}} = \mathbf{C}_{\hat{y},diag} + \mathbf{C}_{\hat{y},off} \tag{48}$$

where the diagonal component is simply the diagonal matrix whose elements are the diagonal elements of $\mathbf{C}_{\hat{y}}$ and where the off-diagonal component has zeros on the diagonal and coincides elsewhere with the matrix $\mathbf{C}_{\hat{y}}$.

An index which assesses the dominance of the diagonal term over that off-diagonal might be simply defined from the norms of the matrices in Eq. (48). However, the norm is not additive. In general, we have

$$\text{norm}(\mathbf{C}_{\hat{y}}) \neq \text{norm}(\mathbf{C}_{\hat{y},diag}) + \text{norm}(\mathbf{C}_{\hat{y},off}) \tag{49}$$

therefore, an index such as the ratio of the $\mathbf{C}_{\hat{y},diag}$-norm to the $\mathbf{C}_{\hat{y}}$-norm is not well defined. It could be less or greater than one depending on the given matrix.

To quantify the relative contribution of the diagonal term (and off-diagonal term) to the norm of $\mathbf{C}_{\hat{y}}$, let us consider the SVD decomposition of $\mathbf{C}_{\hat{y}}$. With the usual notation, we have

$$\mathbf{C}_{\hat{y}} = \mathbf{C}_{\hat{y},diag} + \mathbf{C}_{\hat{y},off} = \mathbf{U}\mathbf{S}_{\hat{y}}\mathbf{V}^t \tag{50}$$

from which

$$\mathbf{U}^t\mathbf{C}_{\hat{y},diag}\mathbf{V} + \mathbf{U}^t\mathbf{C}_{\hat{y},off}\mathbf{V} = \mathbf{S}_{\hat{y}} \tag{51}$$

which with the position

$$\begin{aligned} \mathbf{B}_{\hat{y},diag} &= \mathbf{U}^t\mathbf{C}_{\hat{y},diag}\mathbf{V} \\ \mathbf{B}_{\hat{y},off} &= \mathbf{U}^t\mathbf{C}_{\hat{y},off}\mathbf{V} \end{aligned} \tag{52}$$

gives

$$\mathbf{B}_{\hat{y},diag} + \mathbf{B}_{\hat{y},off} = \mathbf{S}_{\hat{y}} \tag{53}$$

Finally, because of the definition of norm, we have

$$\text{norm}(\mathbf{C}_{\hat{y}}) = \mathbf{S}_{\hat{y}}(1,1) = \mathbf{B}_{\hat{y},diag}(1,1) + \mathbf{B}_{\hat{y},off}(1,1) \tag{54}$$

where $\mathbf{X}(1,1)$ is the element (1,1) of the given matrix $\mathbf{X}$. The formula above is fully additive and allows us to decompose the norm of the retrieval covariance matrix in its diagonal and off-diagonal contribution. Thus, a proper index which quantify the degrees of diagonalization of $\mathbf{C}_{\hat{y}}$, that is how much the matrix is dominated by its diagonal terms, can be defined by

$$i_D = \frac{\mathbf{B}_{\hat{y},diag}(1,1)}{\mathbf{S}_{\hat{y}}(1,1)} \tag{55}$$

For a full diagonal matrix, we have $i_D = 1$ and the retrieval is truly independent, whereas for a highly correlated matrix, we have $i_D = M^{-1}$. The index (55) can be easily re-scaled to the range 1 to M by simply redefining it as

$$i_D = M \frac{\mathbf{B}_{\hat{y},diag}(1,1)}{\mathbf{S}_{\hat{y}}(1,1)} \tag{56}$$

Then, $i_D = 1$ simply means that for the retrieval at hand it is as if the full atmosphere had been divided just in one layer, that is only the columnar amount of the parameter has been resolved. On the opposite edge of the i_D scale, we have $i_D = M$, and the retrieval has been fully resolved on the grid mesh used to divide the atmosphere. Nearby layers can then, e.g., be used to form average quantities and, therefore, reduce the estimation error.

4. Implementation with Simulated Data and Assessment of the Retrieval Performance

For illustrative purposes the regression methodology outlined above has been implemented by using:

- An ECMWF-analyses-derived training data-set of tropical profiles [6] for clear-sky and sea surface, to build up the orthogonal basis and estimate the regression coefficients. The data-base consists of $n = 377$ diverse profiles for temperature, water vapor and ozone, together with surface temperature.
- the σ-IASI code ([1]) for the computation of synthetic spectral radiance.

The ECMWF (European Center for Medium range Weather Forecasts) data set has been specifically designed to provide a limited number of cases, but representative of the wide variability of the atmosphere, for purposes such as statistical regression.

The regression scheme has been implemented by considering nadir-view NAST-I-like spectra. Therefore, the spectral parameters relevant to the ER-2 NAST-I were used to yield synthetic radiance (basically, the Instrumental Response Function was assumed to be a sine cardinal with a sampling rate of 0.2411 cm^{-1}). The observational covariance matrix, $\mathbf{C}_o$ was assumed to be diagonal and build up on the basis of the NAST-I radiometric noise shown in Fig. (3). Furthermore, it was assumed that NAST-I flew (on board the ER-2) at a flight altitude of $\approx$ 20 km, which corresponds to an atmospheric pressure of about 55 mbar, which in turns corresponds to 40 σ-IASI layers. Thus, the size of a given atmospheric parameter vector is $M = 40$.

For the analysis here shown the following spectral ranges were considered for the EOF regression,

- 700 to 810 cm^{-1}
- 1010 to 1080 cm^{-1}
- 1100 to 1200 cm^{-1}
- 1450 to 1550 cm^{-1}

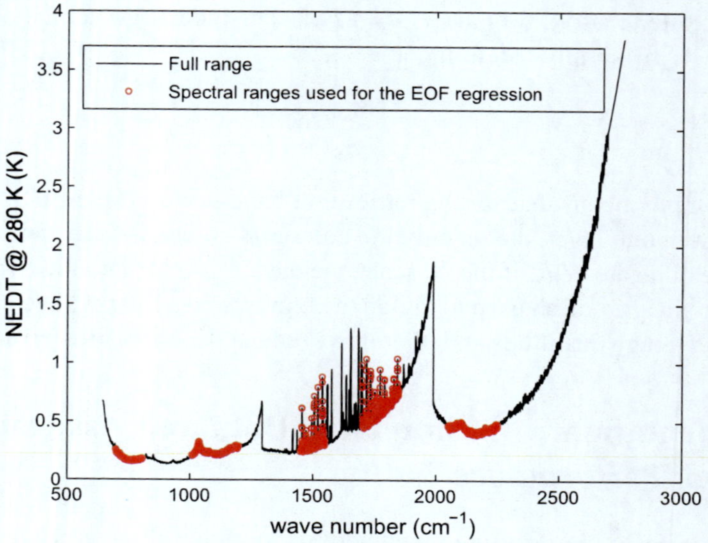

Figure 3. NAST-I radiometric noise in terms of Noise Equivalent Difference Temperature (NEDT) at a scene temperature of 280 K. The figure also evidences the spectral ranges which were used for the EOF regression. These noise values have to be properly transformed to radiance units before using them to build up the covariance matrix, $\mathbf{C}_o$.

- 1700 to 1850 cm^{-1}
- 2050 to 2250 cm^{-1}

These spectral intervals have been chosen in such a way that they correspond to the spectral regions where NAST-I has its best radiometric performance (see Fig. (3)), with the condition that they have to suitably constrain the retrieval for temperature, water vapor and ozone. They contain, indeed, the most fundamental absorption bands for CO_2 (used for temperature), H_2O and O_3, respectively. With the choice above, we consider a total of $N = 2926$ data points from each NAST-I spectrum, therefore the size of the radiance vector is $N = 2926$. The related covariance matrix for the training data set has, therefore, a size of $N \times N$. Singular value decomposition of this matrix, allows us to compute the basis $\mathbf{U}$ of orthogonal empirical functions, and, ultimately, to calculate the PC scores through Eq. (4). The singular values of the decomposition also allow us to check the redundancy of the radiance.

This is done by plotting the singular values σ_j^2 vs j as shown in Fig. (4). It is possible to see that σ_j^2 goes very rapidly to zero as j becomes large. Applying the Kaiser criterion introduced in section 3.2., we have that only the first 28 eigenvalues are greater than one. This is to say that only very few PCs (below 30) are needed in the truncated expansion (7) to reproduce fairly well a vector of size $N = 2926$. This is exemplified in the three figures (5) to (7), which compare a) a given (noiseless) signal to the (noisy) observation, b) the signal to the EOF representation of the observation with a truncation level, $r = 10$, and c) the signal to the EOF representation of the observation with a truncation level of $r = 20$,

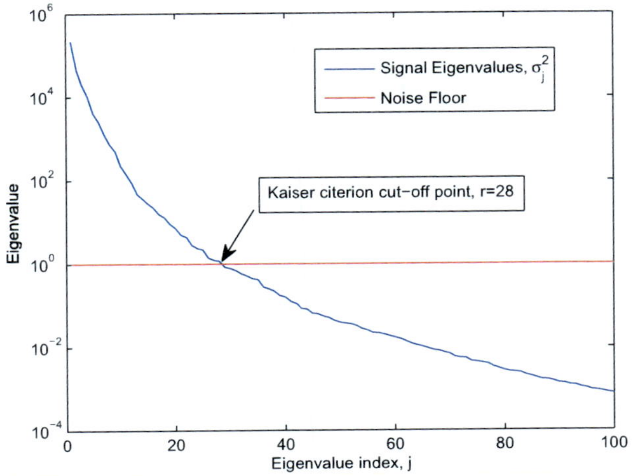

Figure 4. Plot of the first 100 singular values for the spectral radiance *training data set* considered in this study.

respectively. Here the signal represents a typical tropical spectrum in the range 700 to 810 cm^{-1}, which is the core of the CO_2 absorption band. The observation is the *signal + noise,* where the noise is assumed to be Gaussian, zero mean and standard deviation depending on the wave number as shown in Fig. (3).

As expected, the difference *observation-signal* (see Fig. (5) just gives the noise component, which oscillates within the $\pm 1\sigma$ interval. The latter is obtained by the NAST-I radiometric noise shown in Fig. (3). In contrast, the difference in Fig. (6) shows that the $r = 10$-truncated representation of the observation is much closer to the signal that the observation itself. This is due to the noise-filtering of the EOF-truncated expansion. The effect of noise-filtering is much more evident if we consider $r = 20$, as shown in Fig. (7). It should be stressed here that according to the Kaiser criterion there is no advantage in using more than $\approx 20 - 30$ PC scores to represent the data. Higher truncation points would better fit the observation, that is the noisy signal, rather than the signal itself.

Based on these results, the advantage of using a set of PC scores, instead of the high dimensional radiance vector, to implement a suitable regression relation for the atmospheric parameters appears quite obvious. Instead of using a potential of $N = 2926$ *predictors,*we can use, without much loss of information, a few PC scores (10 to 20).

For the *predictants,* that is the atmospheric parameters, we consider a total of $3 \cdot M + 1$ parameters, with $M = 40$. They are:

- $M + 1$ parameters for temperature (the vertical profile plus the surface temperature),
- M parameters for the H_2O vertical profile,
- M parameters for the O_3 vertical profile.

For each of these parameters, the linear regression (15) can be considered and the regression coefficients computed, according to the methodology we have outlined in the pre-

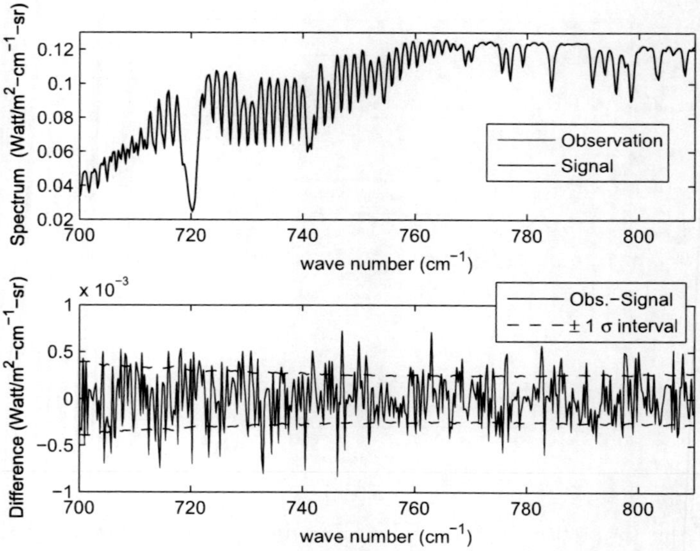

Figure 5. Example of one element of the training data set. Only the spectral interval 700 to 810 cm^{-1} is shown for the sake of clearness. The figure shows both signal (noiseless spectrum) and observation (signal+noise).

vious sections.

4.1. More on "How Many Components do We Need to Extract?"

To be implemented, the above procedure needs that we choose a suitable truncation point, r. One suitable choice might be that dictated by the Kaiser criterion, which in this case, according to the eigenvalues plot shown in Fig. (4), would lead us to select $r = 28$.

However, as already pointed out at the end of section (3.1.), our objective is not to best reproduce a given signal with a limited number of principal components, when rather to find a representation which is capable to minimize the *forecast error* in a problem of linear regression, therefore the strategy we exploit is to choose that r for which the regression or forecast error, $E\left((y-\hat{y})^2\right)$ is minimized.

Let $y_{T,j}(i)$ be the temperature corresponding to the i-th layer from the j-th sample for the training data set, $i = 1, \ldots, M+1$ (with $M = 40$) and $j = 1, \ldots, n$, with $n = 377$. Let $\hat{y}_{T,j}(i;r)$ be the estimated value of $y_{T,j}(i)$ from the EOF regression corresponding to r PC radiance scores. The dependence of the root mean square error of the EOF regression, hence its performance, on the number, r of PC scores can be analyzed by considering the integrated estimation error

$$S_T(r) = \frac{1}{M \cdot n} \sum_{i=1}^{M+1} \sum_{j=1}^{n} \left(\hat{y}_{T,j}(i;r) - y_{T,j}(i)\right)^2 \tag{57}$$

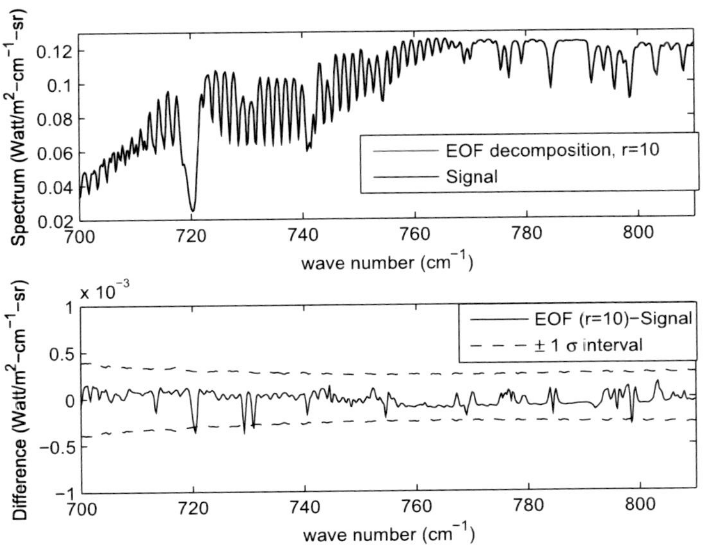

Figure 6. Exemplifying the noise filtering properties of the EOF decomposition. For the same signal and observation shown in Fig. (5), the figure now compares the EOF-truncated expansion ($r = 10$) of the observation to the signal. Compare with Fig. (5).

Similarly, for water vapor and ozone we have

$$S_{H_2O}(r) = \frac{1}{M \cdot n} \sum_{i=1}^{M} \sum_{j=1}^{n} \left(\hat{y}_{H_2O,j}(i;r) - y_{H_2O,j}(i) \right)^2 \tag{58}$$

and

$$S_{O_3}(r) = \frac{1}{M \cdot n} \sum_{i=1}^{M} \sum_{j=1}^{n} \left(\hat{y}_{O_3,j}(i;r) - y_{O_3,j}(i) \right)^2 \tag{59}$$

These three quantities are plotted for a range of r from 1 to 100 in Fig. (8).

To have a proper comparison among the three curves, they have been normalized to their corresponding maximum value. From this figure it is possible to see that the regression error reaches a plateau around $r = 50$, that is beyond this limit no further improvement has to be expected in the retrieval performance. However, note that for $r \approx 20$, we have almost the same integrated error as that seen for the plateau. In other words, a value $r = 20$ gives almost the same performance as that for $r \geq 50$. These results are consistent with those shown in the previous three figure, (5) to (7), which demonstrated how a value of $r = 20$ was enough to have a very good reproduction of the signal.

The retrieval error as a function of the altitude (pressure) for $r = 20$ and a value of $r = 100$ which falls well within the plateau is illustrated in Fig. (9). It can be seen that going from $r = 20$ to $r = 100$ only gives a very modest improvement in the performance.

We want to stress that a relative high value for r is not recommended, also in view of the fact that the Kaiser criterion gives $r = 28$. One might believe that by choosing r very deeply in the plateau of Fig. 8 would give us more confidence for a good retrieval

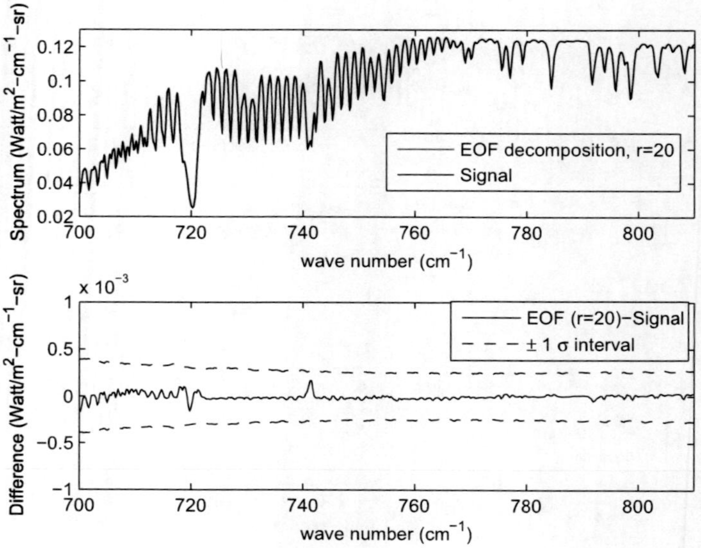

Figure 7. As Fig. (6), but now $r = 20$.

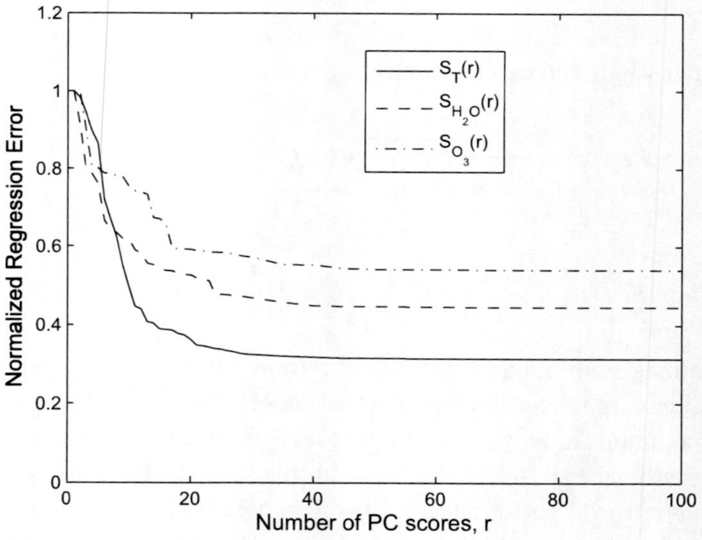

Figure 8. Integrated regression error for temperature, water vapor and ozone as a function of the number of PC scores, r. The error has been normalized to its maximum value.

performance. And in fact, the results shown in Fig. (9) could lead us to such a wrong conclusion. However, in a real situations, principal components, beyond the Kaiser point $r = 28$, are swamped out by noise, so that their inclusion would only add bias to the final results.

A better choice of the truncation point, r implies a direct test on the $S_X(r)$ curves

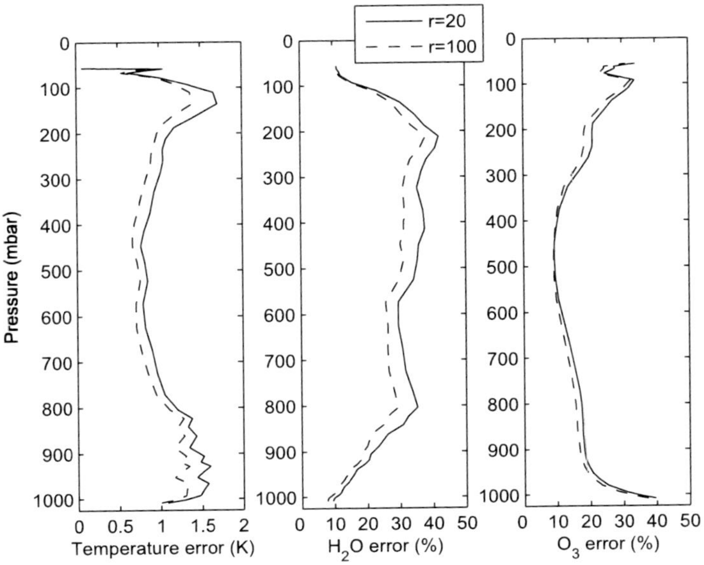

Figure 9. Expected performance for temperature, water vapor and ozone as a function of the altitude for two different values of r.

shown, e.g., in Fig. (8). Here X indicates any of the parameters we are dealing with. In order to select a number of scores, which is close to the plateau, but below the Kaiser criterion point, we can resort to a form of the *scree* test introduced by Cattell [5]. Cattell's recommendation is to retain only those components above the point of inflection on a plot of eigenvalues ordered by diminishing size. Because our specific problem pertains to regression, we can substitute the eigenvalues plot with the $S_X(r)$ curve shown in Fig. (8). To make the Cattell's test suitable of analytical treatment, we define the inflection point in the curve, $S_X(r)$ as the point of maximum curvature. This definition, e.g., is that used to derive the well-known L-curve criterion developed by Hansen [14].

For a generic curve $y = f(x)$, the curvature, k is defined by

$$k = \frac{\mid \frac{d^2y}{dx^2} \mid}{(1 + (\frac{dy}{dx})^2)^{3/2}} \tag{60}$$

To apply the above formula to any single $S_X(r)$ curve, we might perform finite-difference derivative computations. However, this kind of curves are normally very well fitted with algebraic analytic functions of the type

$$\frac{S_X(r)}{S_X(1)} = a + \frac{1-a}{1+(\frac{r}{b})^n} \tag{61}$$

where a, b, n are parameters to be fitted to the data. Once we have the best fitted curve to the data points, we can use it to analytically compute the curvature k and find its maximum.

For the test case shown in Fig. (8), Fig. (10) shows the best fit to the data for the model (61), while Fig. (11) exemplifies for the case of water vapor the curvature curve. According to this plot we have that the optimal choice for r is $r_{opt} = 14$. It is interesting to see that r_{opt} is lower than the Kaiser point, $r = 28$. In other words, the methodology here exemplified for the choice of the number of scores select a point which is both close to the region of best retrieval performance and lower than the Kaiser point, so that allowing us to work with principal components which are well above the noise level. Computing the curvature for the case of temperature and ozone, we have $r_{opt} = 12$ (temperature) and $r_{opt} = 25$, once again both lower than the Kaiser point.

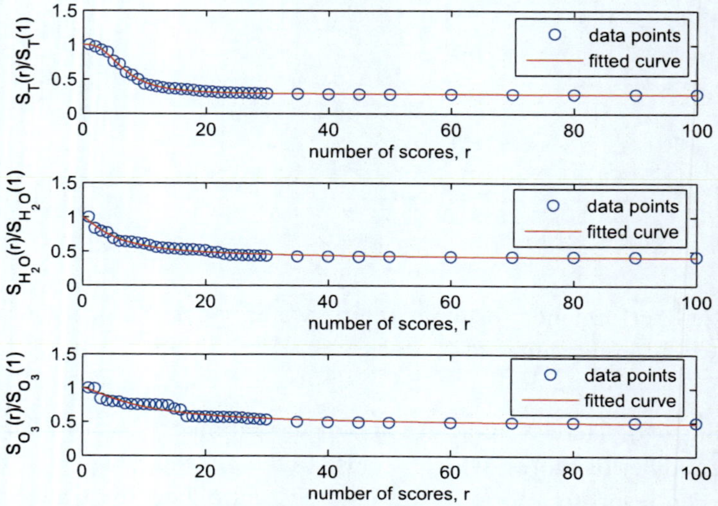

Figure 10. Algebraic model fitted to the integrated retrieval root mean square error, $S_X(r)$.

4.2. Values of the Retrieval Interdependency Index, i_D

Finally, by using the methodology outlined in section 3.3.2., we can analyze the degree of interdependency of the retrieval for temperature, water vapor and ozone. For $r = r_{opt}$, the value of i_D is 5.2 for temperature, 6.4 for water vapor and 2.22 for ozone. These are very modest values! For ozone, $i_D = 2.20$ indicates that only the entire atmospheric column is potentially resolved. For temperature and water vapor, i_D says that only the very coarse features of the profile can be resolved. This is not a serious shortcoming for temperature, which is typically a smooth function of the altitude, but could become a serious limitations for water vapor, whose profile may be characterized by small-scale vertical structures.

5. Application to Real Observations

In this section we discuss the application of the EOF regression methodology to IASI and two different sets of NAST-I observations. To begin with we consider a set of NAST-I data, which has been derived from the CAMEX/3 experiment.

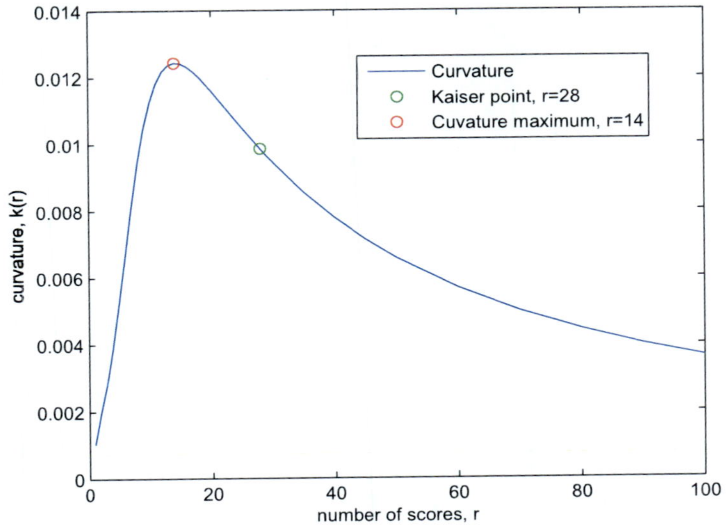

Figure 11. Curvature of the curve, $S_{H_2O}(r)$ obtained with the help of the fitted algebraic model shown in Fig. (10). The figure also shows the position of the Kaiser point and the relative maximum of the curvature itself.

5.1. CAMEX/3 Experiment

During this experiment, NAST-I was on board the ER-2 airplane and was flown at an altitude of $\approx$ 20 km over the radiosonde station (located at 24.70 N Latitude and -77.77 E Longitude) which provided the data needed for the validation of the retrieved products. Once the truth profile (i.e. the radiosonde observation) is known, the inverted products can be directly compared to it and the retrieval accuracy can be assessed.

The CAMEX/3 target area (see Fig. 12) covered a box of size 2.5 degrees times 2.5 degrees, in terms of Latitude/Longitude coordinates. This corresponds to a square of size 280×280 km^2, which is an area for which a certain degrees of atmospheric variability has to be expected.

The NAST-I spectra were processed for clear sky and a number of $n = 294$ spectral observations at nadir view were selected for the retrieval of skin temperature, temperature, water vapor and ozone profile. These soundings cover almost uniformly the target area.

The set of regression coefficients was the same as that derived in the previous section to illustrate the implementation of the method. Following the procedure discussed in section 4.1., for the retrieval examples here shown, the values of $r = 12, 14, 25$ were used for temperature, water vapor and ozone, respectively.

Figure (13) shows the retrieved temperature profile for the full set of 294 NAST-I spectra. In the same figure, a comparison with the best co-located radiosonde observation is shown, as well. It is seen that the radiosonde observation fully agrees with the retrieval whose variability, as expected in view of the performance shown in Fig. (9), is mostly pronounced in the lower part of the atmosphere. The quality of the retrieval can be also appreciated from Fig. (14), where now the comparison is shown between the radio sound-

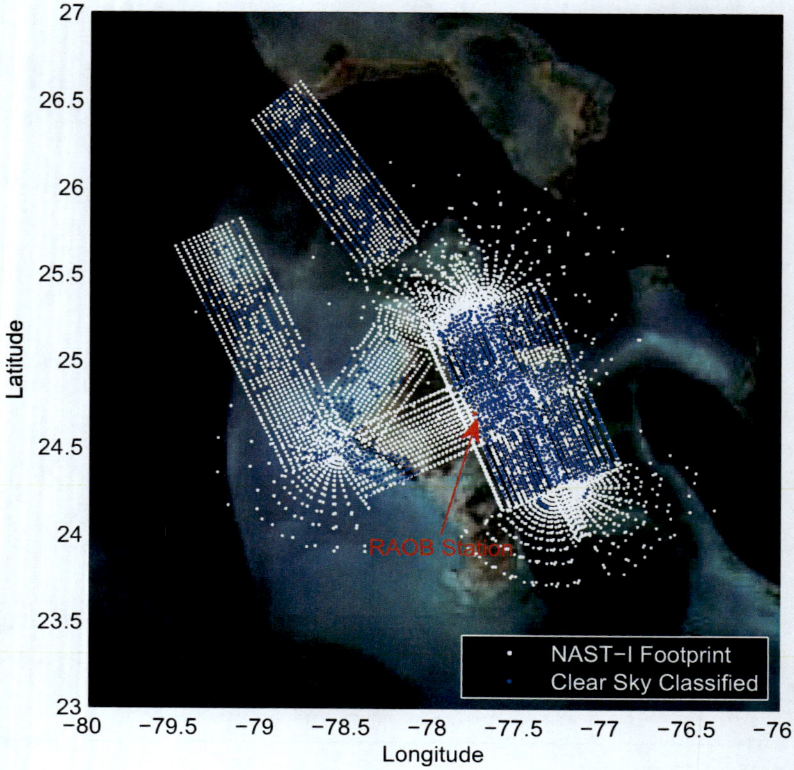

Figure 12. Camex/3 target area showing the NAST-I footprints and those classified as clear sky.

ing and the mean retrieved profile obtained by averaging over the 294 individual inverted profiles. It is seen that differences are at most of 1-2 K and are mostly localized at the level of the tropopause.

The same analysis as that illustrated above for temperature is shown for water vapor in the two figures (15) and (16). From Fig. (15) it is seen that the variability for water vapor is much larger than that exhibited by temperature (see Fig. (13)). This is in part expected in view of the performance shown in Fig. (9). However, the larger variability is also a result of the larger dynamics of H_2O in comparison to temperature. Figure (16) shows that the coarse details of the water profile are fairly well reproduced. It is also seen that the structure around 850 mbar is overestimated. However, the retrieval performance in the lower troposphere is around 10% and better.

5.2. EAQUATE Experiment

The second set of NAST-I spectra, which has been used to exemplify the methodology, refers to the European AQUA Thermodynamic Experiment (EAQUATE) and was recorded over the Adriatic sea during the flight on the night of 9-10 September 2004. NAST-I was on board Proteus and flown at an altitude of $\approx$ 15 Km over the target area shown in Fig. (17).

Because of limitation with truth data, only the soundings over the Adriatic sea (see Fig.

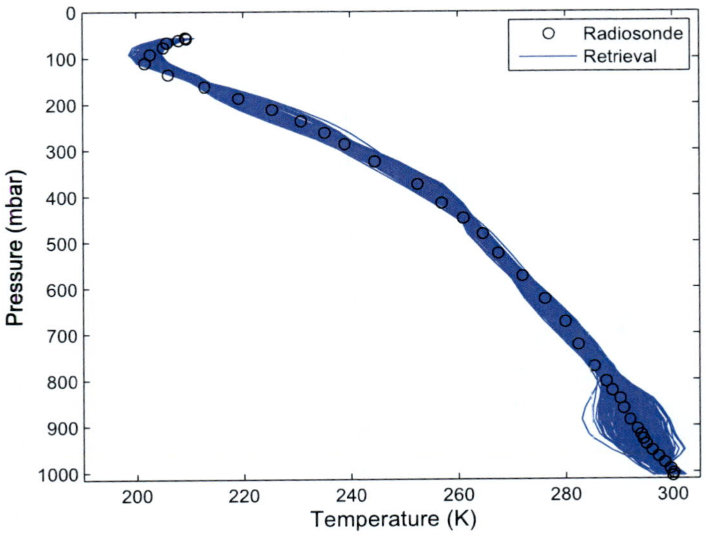

Figure 13. Temperature retrieval for the 294 NAST-I soundings and comparison with the best co-located radiosonde temperature observation. The NAST-I retrievals refer to the CAMEX/3 experiment.

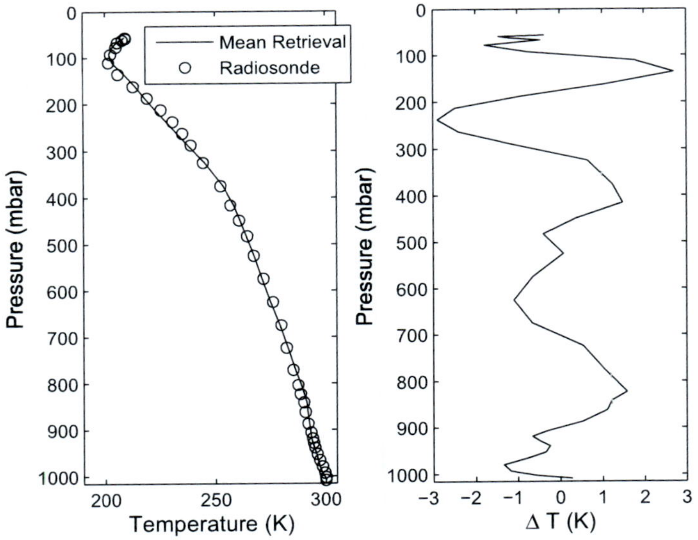

Figure 14. Left panel: mean temperature retrieval obtained by averaging over the 294 individual profiles shown in Fig. (13) and comparison with the radiosonde temperature observation; right panel: difference between the radiosonde observation and the mean retrieval.

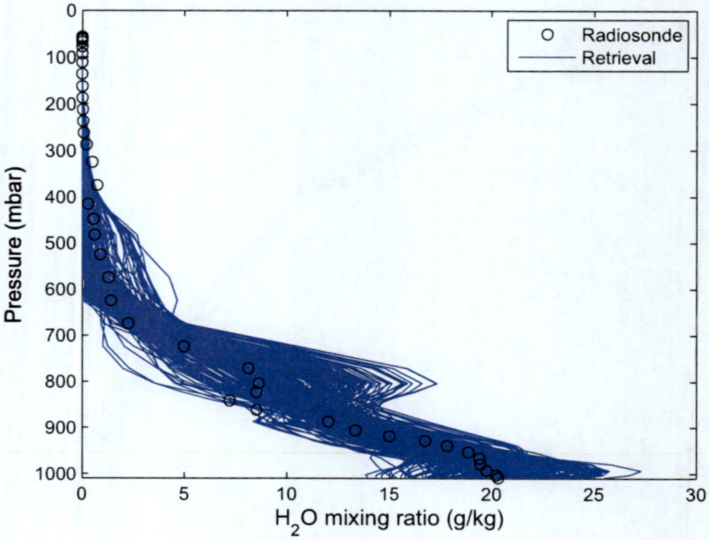

Figure 15. Water vapor retrieval for the 294 NAST-I soundings and comparison with the best co-located radiosonde H_2O observation. The NAST-I retrievals refer to the CAMEX/3 experiment.

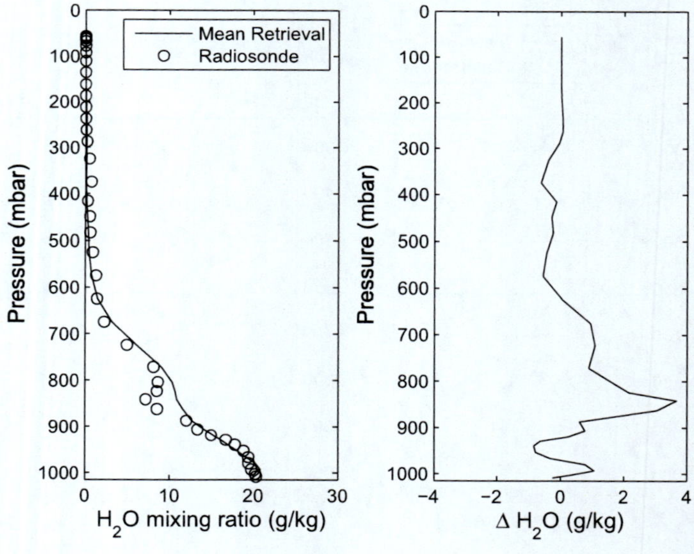

Figure 16. Left panel: mean H_2O retrieval obtained by averaging over the 294 individual profiles shown in Fig. (15) and comparison with the radiosonde H_2O observation; right panel: difference between the radiosonde observation and the mean retrieval.

(17) have been considered for the present analysis. The target area was therefore much more limited in size than that for the CAMEX experiment. It extends along a line of $\approx$ 75 km in

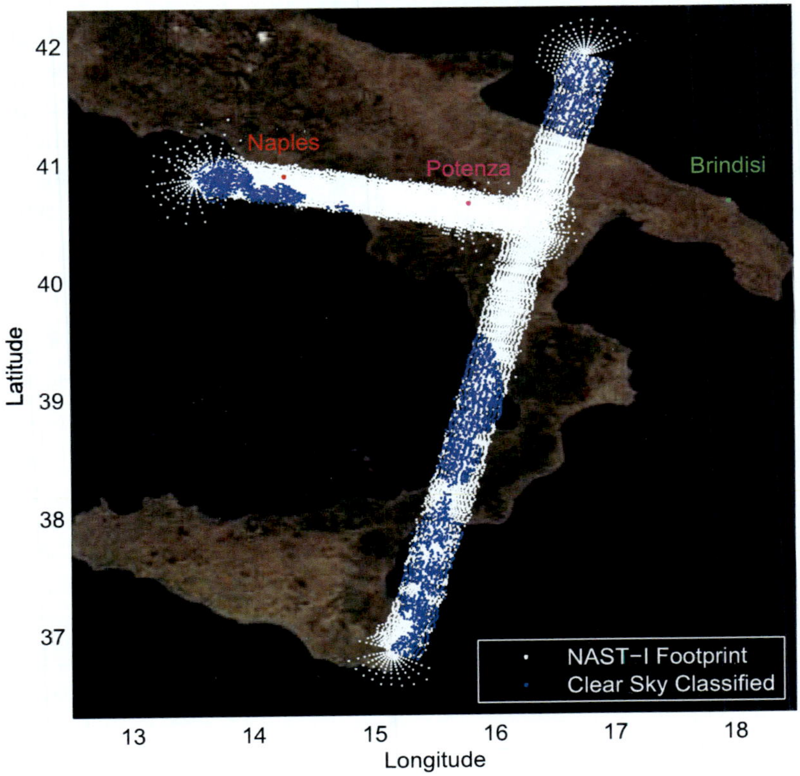

Figure 17. Eaquate target area showing the NAST-I footprints and those classified as clear sky. Only, the soundings over the Adriatic sea have been considered in the analysis.

the direction of the Proteus flight, and its centroid coordinates are 41.50° N, 16.70° E. In contrast to the CAMEX/3 experiment, the radiosonde station was not perfectly co-located with the soundings. It is located at 40.65° N Latitude and 17.95° E longitude; the station is some 150 km apart from the Proteus target area.

The analysis has been limited to NAST-I soundings at nadir, $\pm 7.5°$ and $\pm 15°$ FOV angles. This limited the number of spectra to be analyzed to a total of 270 spectra. The cloud screening process of these 270 observations yielded a series of 46 sea-surface clear-sky NAST-I spectra, which were further considered for the EOF regression retrieval of the skin temperature, and vertical profile of temperature, water vapor and ozone.

We had to redo the training of the EOF regression and to recompute the regression coefficients to take into account for the diverse geographical area with respect to the CAMEX/3 experiment. The new training was performed again by using the Chevalier data base [6]. Only state vectors within the extra tropical North and South 30° to 45° latitudinal belts were considered to derive the EOF regression coefficients. Also note that we needed a set of coefficients for each NAST-I view angles considered in the analysis. As before, for this second example, the truncation point, r was selected using the procedure outlined in section 4.1..

The temperature results for the full set of 46 soundings are shown in Fig. (18), which also shows the comparison with the radiosonde observation. It is seen that the retrieval is fully consistent with the radio sounding for temperature. Differences (at most of ≈ 2 K) are seen in the middle of the troposphere and at the level of the tropopause. The differences are better analyzed by considering Fig. (19), where now the comparison is shown between the radio sounding and the mean retrieved profile obtained by averaging over the 46 individual regressed profiles. Again it is seen that differences are at most of 2 K and are mostly localized in the middle troposphere.

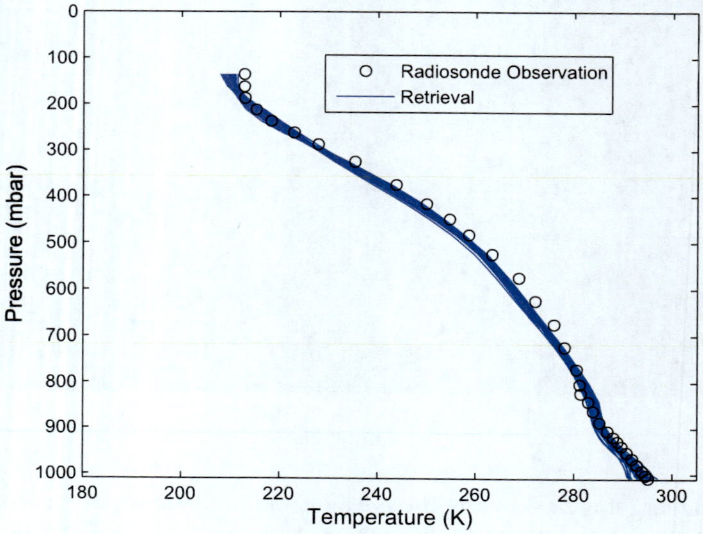

Figure 18. Temperature retrieval for the 46 NAST-I soundings and comparison with the best co-located radiosonde temperature observation. The NAST-I retrievals refer to the EAQUATE experiment.

The water vapor results for the full set of 46 NAST-I soundings are shown in Fig. (20), which also shows the comparison with the radiosonde observation. It is seen that the radiosonde observation falls well within the variability of the set of regressed profiles. Although differences are clearly visible, the retrieval exhibits the same structure as that shown by the sonde observation. Figure (21), which compares the mean retrieval H_2O profile to the radiosonde observation, confirms the above conclusion, even though differences are seen, mostly at the level of 750 mbar, the regressed profiles is within $\approx 10\%$ in the range 800 to 1000 mbar.

5.3. IASI Tropical Soundings

Finally, we apply the EOF regression scheme to a set of IASI soundings over the tropical belt. As said before, the IASI data were acquired during the IASI commissioning phase on 22 July 2007. The geo-location of these soundings can be seen in Fig. (22). In contrast to airplane NAST-I data, the IASI data have been observed from satellite and therefore they potentially bring information over the entire atmospheric column, extending from sea level

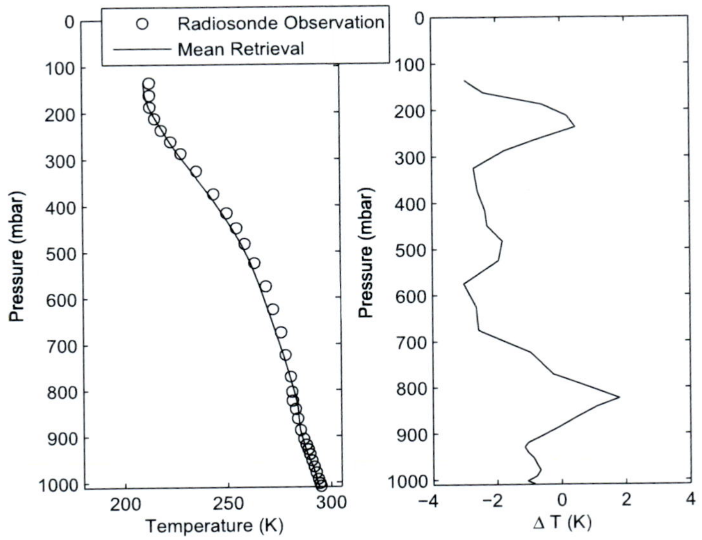

Figure 19. Left panel: mean temperature retrieval obtained by averaging over the 46 individual profiles shown in Fig. (18) and comparison with radiosonde temperature observation; right panel: difference between the radiosonde observation and the mean retrieval.

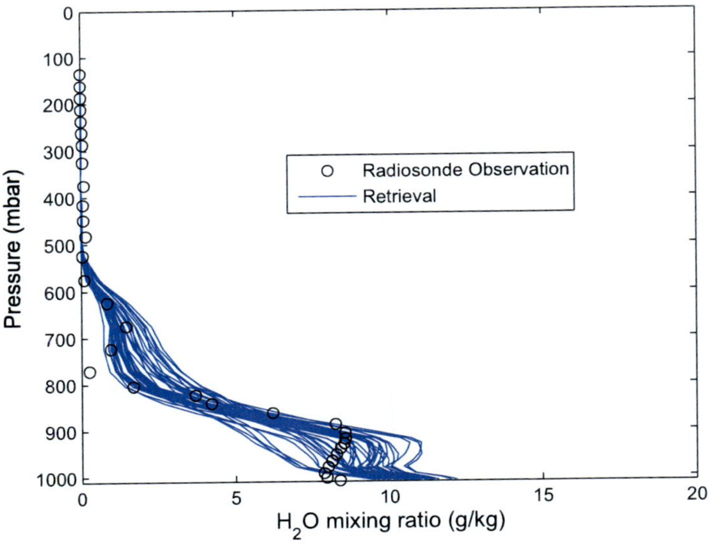

Figure 20. Water vapor retrieval for the 46 NAST-I soundings and comparison with the best co-located radiosonde H_2O observation. The NAST-I retrievals refer to the EAQUATE experiment.

(about 1000 mbar) to the top of the atmosphere.

In total, a number of 603 IASI spectra has been selected. The spectra have been ob-

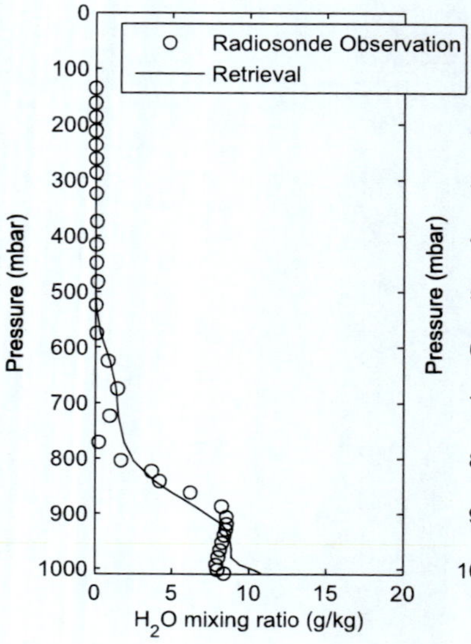

Figure 21. Left panel: mean H_2O retrieval obtained by averaging over the 46 individual profiles shown in Fig. (20) and comparison with radiosonde H_2O observation; right panel: difference between the radiosonde observation and the mean retrieval.

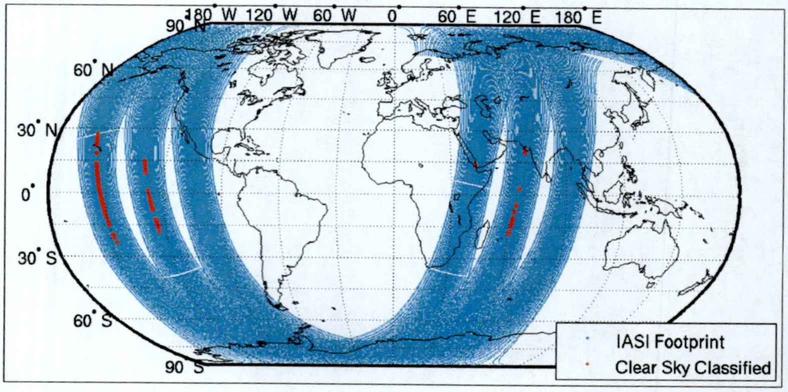

Figure 22. IASI orbits for the day 22 July 2007 and clear sky footprints (in red) considered in this analysis.

served on sea surface and refer to clear sky conditions. To simplify the illustration of the methodology only nadir view soundings have been considered.

To develop a consistent set of truth data against which IASI retrieval could be compared, ECMWF atmospheric analysis fields for temperature, water vapor and ozone were considered. These fields where time and spatially co-located to the 603 IASI soundings. We used atmospheric analysis fields of 00:00, 06:00, 12:00 18:00 and 24:00 UTC on 22 July 2007. A detailed description of the accuracy and horizontal spatial resolution of ECMWF analysis fields can be found in [13].

To apply the EOF regression scheme to IASI case, we had to redo the training of the EOF regression and to recompute the regression coefficients to take into account for the diverse platform (satellite vs airplane) and the different spectral characteristics between IASI and NAST-I (that is Instrumental Spectral Response Function and radiometric noise). The new training was performed again by using the Chevalier data base [6]. Only state vectors within the tropical belt, $\pm 30°$ latitude were considered to derive the EOF regression coefficients. Only nadir view soundings were considered. As before, for this third example, the truncation point, r was selected using the procedure outlined in section 4.1..

Figure (23) compares the IASI retrieved skin temperature to the corresponding ECMWF analysis.

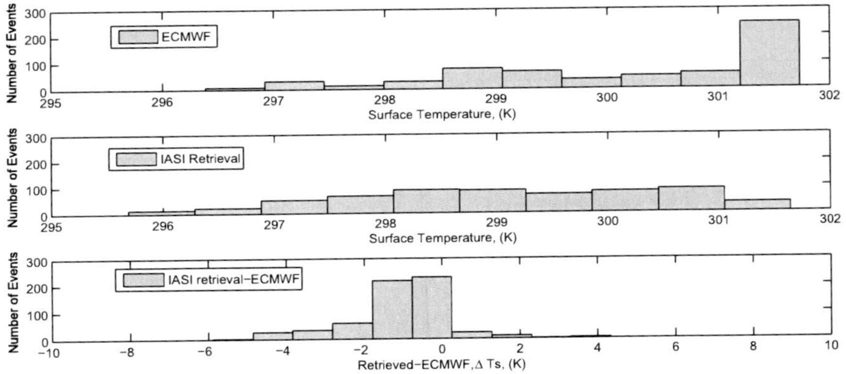

Figure 23. Histogram of IASI retrieval for the sea surface temperature compared with that for the ECMWF analysis.

It is possible to see that IASI senses, on average, a slightly colder surface. The mean difference is about 1 K, whereas the root mean square difference is 1.2 K. This behavior is due to the smooth nature of the ECMWF analysis once compared to IASI data, whose horizontal resolution, at nadir, is about 12 km.

The temperature results for the full set of 603 soundings are shown in Fig. (24), which also shows the comparison with the ECMWF analysis. It is seen that the retrieval is fully consistent with the ECMWF analysis for temperature. Differences (at most of $\approx$ 2 K) are seen in the middle of the troposphere and at the level of the tropopause. The differences are better analyzed by considering Fig. (25), where now the comparison is shown between the mean values of analysis and retrieval. Of course this mean values have been obtained by averaging over the 603 individual regressed and ECMWF profiles, respectively. Again it is seen that differences are at most of 2 K and are mostly localized in the upper part of the atmosphere.

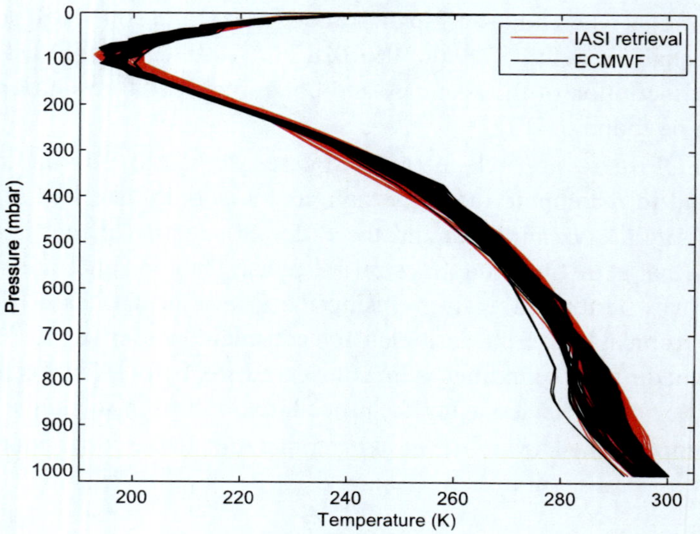

Figure 24. Temperature retrieval for the 603 IASI soundings and comparison with the ECMWF analysis.

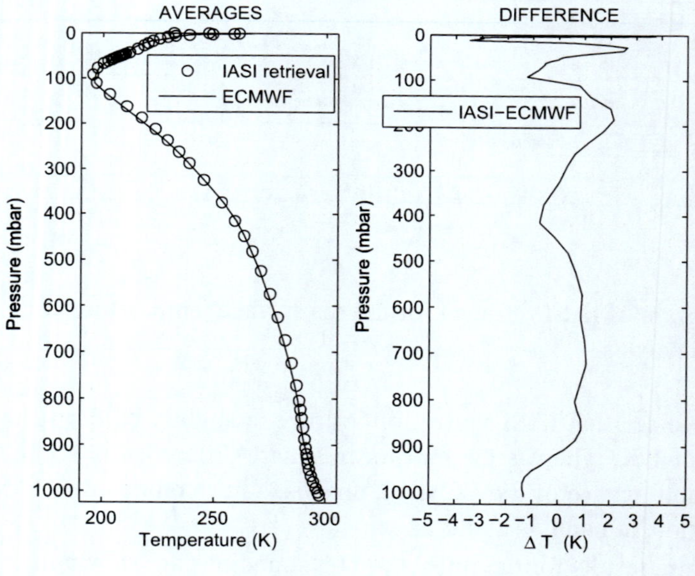

Figure 25. Left panel: mean temperature retrieval obtained by averaging over the 603 individual profiles shown in Fig. (18) and comparison with the equivalent mean profile of the ECMWF temperature analysis; right panel: difference between the ECMWF mean profile and the IASI mean retrieval.

The water vapor results for the full set of 603 IASI soundings are shown in Fig. (26), which also shows the comparison with the ECMWF H_2O analysis. It is seen that the

ECMWF analysis falls well within the variability of the set of regressed profiles. Although differences are clearly visible, the retrieval exhibits the same structure as that shown by the ECMWF filed. Figure (27), which compares the mean retrieved H_2O profile to the ECMWF mean profile, confirms the above conclusion, even though differences are seen, which reach the value of ±1 g/kg in the lower troposphere, the regressed profile is within $\approx$ 10% of that produced by the ECMWF analysis.

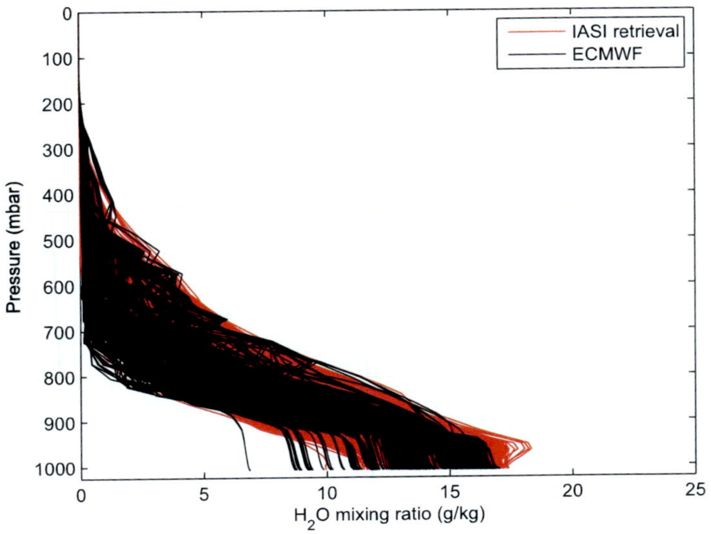

Figure 26. H_2O retrieval for the 603 IASI soundings and comparison with the ECMWF analysis.

For the IASI exercise we can also perform a comparison for ozone retrieval. This is provided in Fig.s (28) and (29). The comparison evidences, once again, the very nice consistency of the IASI retrieval with the ECMWF analysis. In the lower to middle stratosphere, where the bulk of ozone is concentrated, the difference is within 10%.

6. Conclusion

This study has described and demonstrated a methodology for the statistical retrieval of temperature and gas species concentration that uses the spectral radiance measured by the next generation of high-resolution satellite-borne infrared sensors. The performance of the retrieval scheme has been assessed on the basis of numerical exercises. Examples of retrievals based on IASI and NAST-I spectra measured over the sea surface have been discussed, which demonstrate the ability of the scheme to obtain accurate estimation of the geophysical parameters.

The EOF regression methodology has been developed within the context of a generic signal-noise model, and, therefore, our analysis provides a general framework in which to design and develop specific user applications.

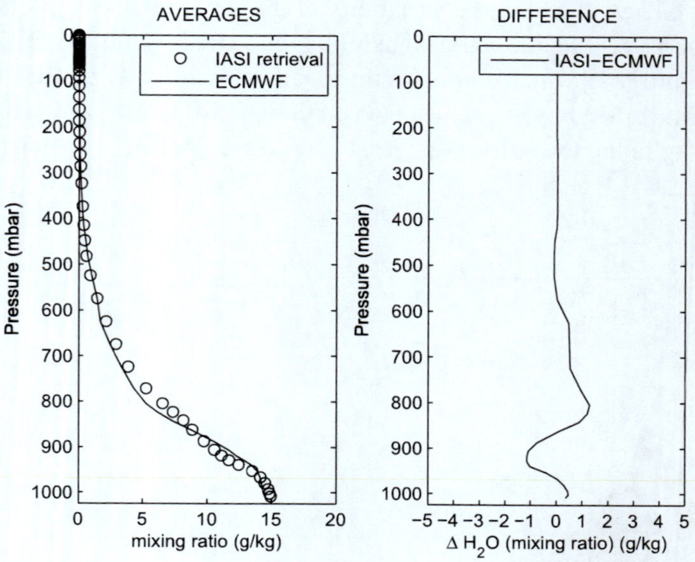

Figure 27. Left panel: mean temperature retrieval obtained by averaging over the 603 individual profiles shown in Fig. (26) and comparison with the equivalent mean profile of the ECMWF H_2O analysis; right panel: difference between the ECMWF mean profile and the IASI mean retrieval.

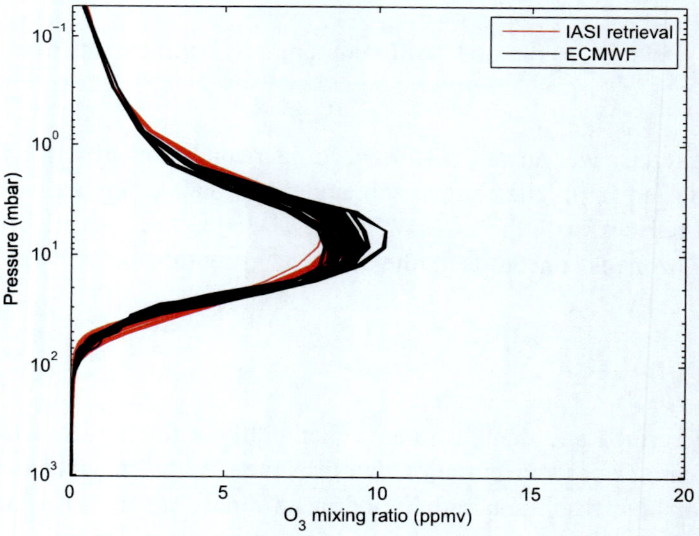

Figure 28. O_3 retrieval for the 603 IASI soundings and comparison with the ECMWF analysis.

The methodology has been exemplified by considering various retrieval examples of atmospheric parameters from high spectral resolution infrared observations. These examples

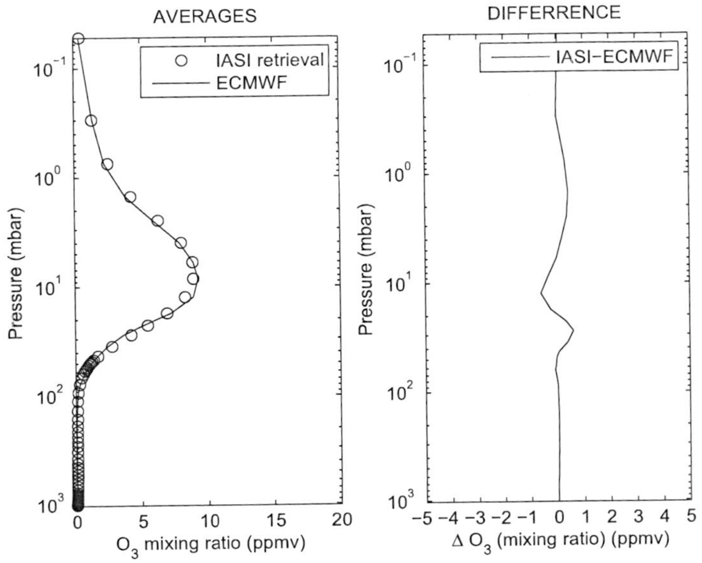

Figure 29. Left panel: mean temperature retrieval obtained by averaging over the 603 individual profiles shown in Fig. (28) and comparison with the equivalent mean profile of the ECMWF O_3 analysis; right panel: difference between the ECMWF mean profile and the IASI mean retrieval.

show that for temperature a retrieval performance of 1-2 K accuracy can be obtained. For water vapor the method yields rather smooth results which does not seem to exactly follow the fine structures present in the vertical profile of this atmospheric constituents. However, the accuracy is within 10% in the lower atmosphere.

In principle the methodology can be used to invert spectral radiance for any atmospheric parameter and we have shown that ozone (mostly columnar amount) can be obtained with a potential good accuracy. Results for ozone have been shown for the case of IASI data.

Our findings, which encompasses a large variety of meteorological situations and different geographical regions, lead us to conclude that the EOF methodology can form the basis for operational schemes for the retrieval of atmospheric parameters. Finally, it can be used to initialize non linear physical inverse schemes, a procedure that the authors have been proved to be particularly successful (e.g. [4, 13]).

Acknowledgment

This work was in part supported by MIUR PRIN 2005 project # 2005025202/Area 02. The author wish to thank the NASA NAST-I team, who provided the data, which were used in this study. We thank Dr M. Matricardi who provided the ECMWF analyses. IASI has been developed and built under the responsibility of the Centre National d'Etudes Spatiales (CNES, France). It is flown onboard the Metop satellites as part of the EUMETSAT Polar System. The IASI L1 data are received through the EUMETCast near real time data distribution service.

References

[1] Amato, U., Masiello, G., Serio, C., and Viggiano, M., The σ-IASI code for the calculation of infrared atmospheric radiance and its derivatives *Environmental Modelling&Software* **17**(7), 651–667 (2002).

[2] Beltrami, E., Sulle funzioni bilineari *Giornale di Matematiche* **11**, 98-106 (1873).

[3] Calbet, X., and Schlüssel, P., Technical note: analytical estimation of the optimal parameters for the EOF retrievals of the IASI level 2 product processing facility and its application using AIRS and ECMWF data *Atmos Chem Phys Discuss* **5**, 9691-9730 (2005).

[4] Carissimo, A., DeFeis, I., and Serio, C., The physical retrieval methodology for IASI: the δ-IASI code *Environmental Modelling&Software* **20**(9), 1111–1126 (2005).

[5] Cattell, R. B., The scree test for the number of factors. *Multivariate Behavioral Research*, **1**, 629-637, (1966).

[6] Chevalier, F., *Sampled databases of 60-level atmospheric profiles from the ECMWF analyses, Research Report No. 4, SAF programme*, EUMETSAT/ECMWF, Reading, UK, (2001).

[7] Chichocki, A., and Unbehauen, R., *Neural Networks for Optimization and Signal Processing*, Wiley, NY, (1996).

[8] Cousins, D., and Gazarick, M.J., NAST Interferometer Design and Characterization. in *Final Report, MIT Lincoln Laboratory Project Report NOAA-26,*July 13, (1999).

[9] Goldberg, M.D., Qu,Y., McMillin, L.M., Wolf, W., Zhou, L., and Divakarla, M., AIRS near-real-time products and algorithms in support of operational numerical weather prediction *IEEE Trans. Geosci. Remote Sensing* **41**, 379-389 (2003).

[10] Golub, G.H., Kahan, W., Calculating the singular values and pseudo-inverse of a matrix *SIAM J. Numer. Anal. Ser. B* **2**, 205224 (1965).

[11] Golub, G.H., Reinsch, C., Singular Value Decomposition and Lesat Squares Solutions *Numer. Mat.* **14**, 403-420 (1970).

[12] Grieco, G., Luchetta, A., Masiello, G., Serio, C., and Viggiano, M., IMG O_3 retrieval and comparison with TOMS/ADEOS columnar ozone: an analysis based on tropical soundings *J. Quant. Spectrosc. Radiat. Transfer* **95**(3), 331-348 (2005).

[13] Grieco, G., Masiello, G., Matricardi, M., Serio, C., Summaa, D., Cuomo, V., Demonstration and validation of the φ-IASI inversion scheme with NAST-I data *Q. J. R. Meteorol. Soc.* **133**(S3), 217-232 (2007).

[14] Hansen, C., Analysis of discrete ill-posed problems by means of the L-curve. *SIAM Review*, **34** (4), 561580 (1992).

[15] Hotelling, H., Analysis of a Complex of Statistical Variables with Principal Components *Journal of Educational Psychology* **24**, 417-441, 498-520, (1933).

[16] Huang, H-L, Antonelli, P., Application of Principal Component Analysis to High-Resolution Infrared Measurements Compression and Retrieval *J. Appl. Met.* **40**, 365-388 (2001).

[17] Jolliffe, I.T., *Principal component analysis,* New York : Springer-Verlag, (2002).

[18] Kaiser, H. F., The application of electronic computers to factor analysis. *Educational and Psychological Measurement*, **20**, 141-151, (1960).

[19] Karhunen, K., *On linear methods in probability theory,* English translation by I. Selin, The Rand Corporation, Doc. T-131, August 11, 1960, 1947.

[20] Kobayashi, H., Shimota, A., Yoshigahara, C., Yoshida, I., Uehara, Y., Kondo, K., Satellite-borne high-resolution FTIR for lower atmosphere sounding and its evaluation *IEEE Trans. Geosci. Remote Sensing* **37**(3), 1496–1507, (1999).

[21] Loéve, M., *Fonctions Aléatoires de Seconde Ordre,* in P. Levy, Processus Stochastiques et Mouvement Brownien, Hermann, Paris, (1948).

[22] Lorenz, E., Empirical orthogonal functions and statistical weather prediction. In *Tech. Rep. 1, Statistical Forecasting Project,* Massachusetts Institute of Technology, Cambridge, MA, 49 pp.

[23] Masiello, G. and Serio, C., Dimensionality-reduction approach to the thermal radiative transfer equation inverse problem *Geophys. Res. Lett.* **31**(11), L11105-L11112 (2004).

[24] Obukhov, A.M., The Statistically Orthogonal Expansion of Empirical Functions, *Izvestiya, Seriya Geojizicheskaya, Akademiya Nauk, SSSR,*No. 3, March 1960, pp. 432-439. English translation by the American Geophysical Union, Nov. 1960, pp. 288-2911.

[25] Pearson, K., On lines and planes of closest fit to systems of points in space *Philosophical Magazine* **2**, 559-572, (1901).

[26] Rao, C.R., The use and interpretation of principal component analysis in applied research *Sankhya Ser. A* **26**, 329-358, (1964).

[27] Serio, C., Esposito, F., Masiello, G., Pavese, G., Calvello, M.R., Grieco, G., Cuomo, V., Buijs, H.L., and Roy, C.B., Interferometer for ground-based observations of emitted spectral radiance from the troposphere: evaluation and retrieval performance, *Appl. Opt.*, **47**(21), 3909-3919, (2008)

[28] Taylor, J.P, Smith, W.L, Cuomo V., Larar, A.M, Zhou, D.K, Serio, C., Maestri, T., Rizzi, R., Newman, S., Antonelli, P., Mango, S., Di Girolamo, P., Espsoito, F., Grieco, G., Summa, D., Restieri, R., Masiello, G., Romano, F., Pappalardo, G., Pavese, G., Mona, L., Amodeo. A., Pisani, G., EAQUATE An International Experiment For Hyper-spectral Atmospheric Sounding Validation *Bulletin of the American Meteorological Society, February issue*, pp. 203-218 doi:10.1175/BAMS-89-2-203 (2008).

[29] Wark, D.G., Fleming, H.E., Indirect Measurements of Atmospheric Temperature Profiles from Satellites: I. Introduction *Monthly Weather Review* **94**(6), 351-362, (1966).

[30] Zhou, D.K., Smith, W.L., Li, J., Howell, H.B., Cantwell, G.W., Larar, A.M., Knuteson, R.O., Tobin, D,C., Revercomb, H.E., Bingham, G.E., Tsou, J-.J., Mango, S.A., Thermodynamic Product Retrieval Methodology and Validation for NAST-I *Appl. Opt.* **41**(33), 6957–6970, (2002).

In: Environmental Modelling: New Research
Editor: Paul N. Findley, pp. 89-102
ISBN: 978-1-60692-034-3

Chapter 3

COMOVEMENT AND CYCLICAL PATTERNS OF SOUTHERN PINE BEETLE OUTBREAKS

Jianbang Gan*
Department of Ecosystem Science and Management, Texas A&M University

Abstract

Insect infestations have been a major driving force of landscape change, leading to severe ecological and economic consequences. The southern pine beetle (SPB), *Dendroctonus frontalis* Zimmermann, is the most destructive insect to pine forests in the U.S. South. This study probes the spatial and temporal patterns, particularly comovement and cyclical patterns, of SPB infestations at broad scales in the Southern United States. Cluster analysis in terms of comovement shows that SPB infestations in the region can be classified into three subregions: Alabama-Florida-Louisiana-Mississippi-Virginia, Georgia-Carolinas-Tennessee, and Arkansas-Texas. SPB infestation risk has increased over time in Florida, South Carolina, and Tennessee, but decreased in Alabama. The magnitude of bi-state comovements of SPB infestations is in general quite large whereas that of regionwide comovements is small, and comovments of small outbreaks are more pronounced than those of big ones. SPB infestations in North Carolina best resemble (synchronize with) the region's median, and thus it can be used as the region's reference for monitoring and forecasting. Though regionwide cyclical outbreak patterns are not detected, statistic evidence of cyclical outbreaks for some states and especially for the identified clusters/subregions is apparent; and the sinusoidal component is commonly concentrated over a range of low frequencies. These results will be of value for monitoring and mitigating SPB outbreaks in the region.

Keywords: Bark beetle; Infestation pattern; Spectral analysis; Cluster analysis; Synchronization; Southern United States.

1. Introduction

The southern pine beetle (SPB), *Dendroctonus frontalis* Zimmermann, represents one of the major threats to the health and productivity of pine forests in the U.S. South. Such threats are

* E-mail address: j-gan@tamu.edu. Phone: (979) 862-4392, fax: (979) 845-6049, Department of Ecosystem Science and Management, 2138 TAMU, Texas A&M University, College Station, TX 77843-2138, USA

likely to intensify with predicted climate change in the future (Gan 2004). This has posed and will continue to pose a tremendous challenge for SPB infestation mitigation. Mitigating SPB impacts depends, to a large extent, upon our understanding and reliable projections of its outbreak patterns.

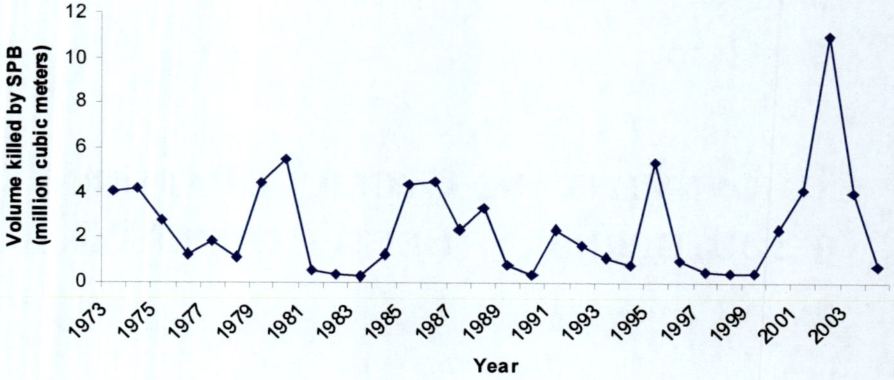

Figure 1. Volumes of sawtimber and pulpwood killed by southern pine beetles in the U.S. South.

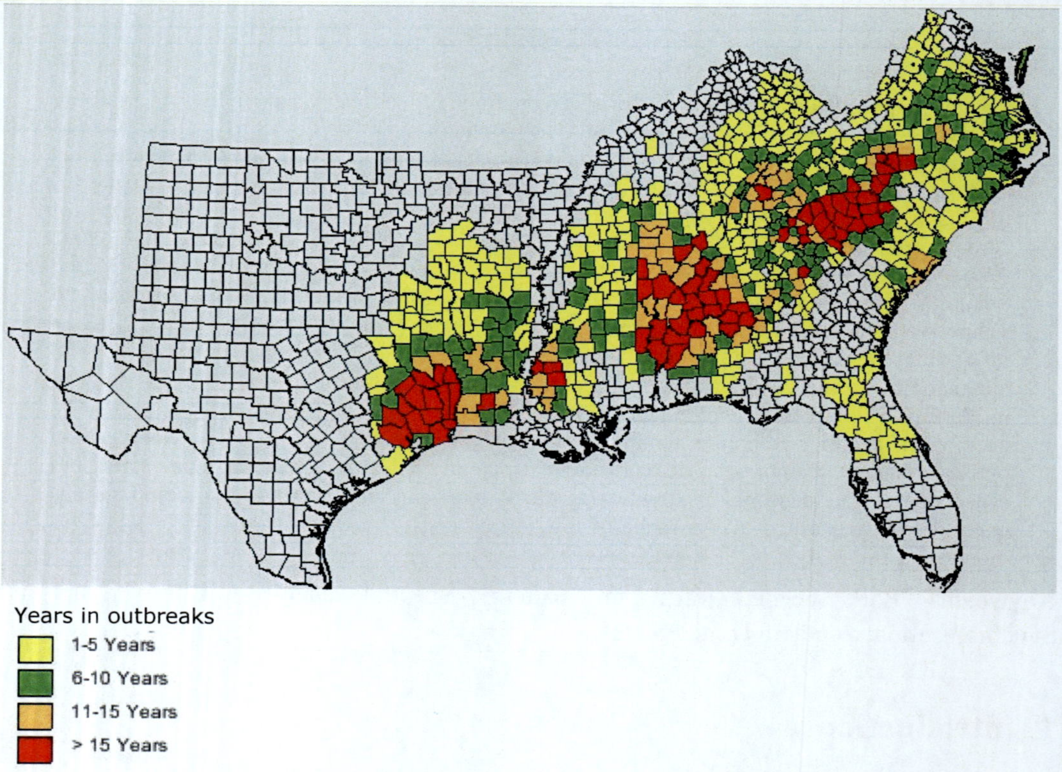

Figure 2. Southern pine beetle outbreaks in the U.S. South, 1960-2004 (map courtesy: Forest Health Protection, Southern Region, USDA Forest Service, http://www.fs.fed.us/r8/foresthealth/atlas/spb/spb.html).

Historical data (Price et al. 1998, Pye et al. 2005) have intuitively shown possible temporal and spatial phenomena of SPB outbreaks in the U.S. South (Figures 1 and 2). Pye (1993) noticed that SPB outbreaks in the Southern United States reoccurred every 6-7 years, whereas Mawby and Gold (1984) reported the outbreak periodicity varied across subregions. Spatial patterns, especially spatial correlations, of SPB outbreaks have also been recognized (Mawby and Gold 1984, Pye 1993, Gumpertz et al. 2000). Yet, in-depth analysis of such patterns is rare and much needed in order to better understand the characteristics of these outbreak patterns and ultimately to guide future SPB monitoring and mitigation efforts.

This study aims at portraying the comovement (spatial) and cyclical (temporal) patterns of SPB infestations at broad scales in the Southern United States. It specifically focuses on addressing the following questions. How closely have SPB outbreaks in different spatial units or subregions moved together? Where are the boundaries of each subregion within which a similar infestation pattern exists? Is there a spatial unit that can serve as the region's reference for SPB infestations? Are there cyclical outbreak patterns? If yes, what are the frequencies or the lengths of cycles? Have the magnitudes of comovement and infestation risk changed over time? A suit of analytical tools, including comovement analysis, cluster analysis, spectral analysis, and others were used to derive answers to these questions and to shed new light on SPB infestation patterns in the region.

The reminder of this article is organized as follows. The next section will describe the measure of SPB infestation risk, methods for modeling comovement and cyclical patterns, and data sources. The results will then be presented and discussed, followed by conclusion.

2. Methods and Data

2.1. Measuring Infestation Risk

The variable used for spectral and comovement analysis is SPB infestation risk, defined as the ratio of the total wood volume killed by SPB to the total volume of pine growing stock. The data on SPB-damaged wood volume were drawn from Pye et al. (2005). Because of missing data for some states, only 11 states were included in this study, including Alabama, Arkansas, Florida, Georgia, Louisiana, Mississippi, North Carolina, Southern Carolina, Tennessee, Texas, and Virginia. The time span of available data varied from state to state though several states have recorded SPB infestations since the early 1960s. All available data were used for spectral analysis, but only the data from 1973 to 2004 were employed for comovement analysis due to the need for balanced data.

The data on the volume of pine growing stock were taken from the Forest Inventory and Analysis (FIA) (Smith et al. 2004, USDA Forest Service 2008). Because forest inventory in the U.S. was conducted at an interval of 5-10 years in the past, linear interpolation was used to derive pine volumes for the intermediate years between two forest inventories. Such an approach is reasonable because the volume of pine growing stock in the U.S. South has had very small variations since the 1950s (Smith et al. 2004).

Table 1. Dickey-Fuller unit root test results of SPB infestation risks.

State	Tau (p-value), with zero mean	Tau (p-value), with drift	Tau (p-value), with trend
Alabama	-3.34 (0.0015)	-4.75 (0.0006)	-5.16 (0.0012)
Arkansas	-2.76 (0.0074)	-3.50 (0.0145)	-3.52 (0.0553)
Florida	-3.24 (0.0021)	-3.62 (0.0108)	-4.20 (0.0123)
Georgia	-2.87 (0.0056)	-4.27 (0.0020)	-4.39 (0.0078)
Louisiana	-3.47 (0.0010)	-3.78 (0.0069)	-3.75 (0.0328)
Mississippi	-2.46 (0.0157)	-3.77 (0.0074)	-3.72 (0.0358)
N. Carolina	-2.93 (0.0043)	-3.85 (0.0049)	-4.00 (0.0163)
S. Carolina	-3.26 (0.0018)	-4.04 (0.0030)	-4.47 (0.0052)
Tennessee	-3.30 (0.0018)	-3.58 (0.0126)	-3.79 (0.0324)
Texas	-3.43 (0.0010)	-4.33 (0.0012)	-4.33 (0.0069)
Virginia	-3.34 (0.0016)	-3.93 (0.0053)	-3.87 (0.0270)

Before performing spectral and comovement analysis, the data series of SPB infestation risks in all spatial units were tested for nonstationarity with the null hypothesis of a unit root using the methods developed by Dickey and Fuller (1979, 1981). No nonstationarity for all the infestation risk data series was detected at the 5% significance level[1] (Table 1). Hence, the infestation risk data series are stationary and can be directly used for modeling without further data manipulation such as differencing or detrending.

2.2. Assessing Comovement

Comovement means "moving with." It is different from correlation and thus cannot be adequately measured by correlation coefficients. For instance, for two series if the standardized shock to one series increases and the other remains unchanged, their correlation coefficient (the product of the standardized residues) will increase though their common movement will become more unequal. There are several approaches that can be used to analyze comovement. One such method is cointegration analysis (Engle and Granger 1987, Murray 1994). Yet, this method is applicable only when all data series in question are integrated of order d, denoted I(d) with $d \neq 0$ and the same for all series. The stationarity, i.e. I(0), of the SPB infestation risk data series (Table 1) precludes the use of cointegration tests.

Baur (2003) developed another approach for measuring and assessing comovement. It has several unique features and does not require the same order of integration in all data series. Following Baur's work, here comovement is defined as the movement of time series that is shared by all series at time t. As such, it can be measured by

$$\Phi_t = \max(s_{1t}, s_{2t}, \ldots, s_{nt})I^- + \min(s_{1t}, s_{2t}, \ldots, s_{nt})I^+ \tag{1}$$

[1] There was only one exception for Arkansas when the test was based the model with a time trend. However, the null hypothesis of nonstationarity for this case was also rejected at the 10% significance level.

where s_{it} is the measure of the feature (e.g. the SPB infestation rate) of the ith series at time t, and I^- (I^+) is a binary variable that is equal to one if all s_{it} for i=1, 2, ..., n are negative (positive) and zero otherwise. Equation (1) is equivalent to

$$\Phi_t = \min[\max(s_{1t}, s_{2t}, ..., s_{nt}), 0] + \max[\min(s_{1t}, s_{2t}, ..., s_{nt}), 0]. \quad (2)$$

Though Φ_t is a simple measure, it suffers from inability of measuring relative movement shared by all series because it is not based on standardized values of s_{it}. For instance, for a given value of Φ_t, one cannot tell exactly the magnitude of the comovement because the means and variances of s_{it} are hidden in this measure. For this reason, relative comovement, defined as the relative movement of time series that is shared by all series at time t, was used in this study. To compute relative comovement, s_{it} needs to be standardized. With the standardized s_{it}, relative comovement can be derived using equation (1) or (2).

Both bivariate and multivariate comovements were analyzed. Bivariate comovements were estimated for all possible pairs of the 11 states. It would also be interesting to know which spatial unit (e.g. state) can serve as a reference for gauging SPB infestations in the region. Such a reference, if existing, will be helpful for monitoring and mitigating SPB outbreaks in the U.S. South. To this end, the median of relative SPB outbreak risks in all the 11 states was chosen as the benchmark. Then, the relative comovement of SPB infestation risk in each state with the region's median was assessed. The state showing the strongest comovement with the region's median may be able to serve as the region's reference for detecting SPB outbreaks.

One of the challenges for multivariate comovement analysis was to determine the proper spatial boundaries of SPB infestations or to properly classify the region into subregions within each of which a similar infestation pattern exists but between which there are distinct patterns. Cluster analysis was performed to identify the states that demonstrated similar temporal SPB infestation patterns. Cluster analysis is a statistic tool for partitioning a set into subsets (clusters) with common features (Everitt et al. 2001). After classification of states, multivariate comovement of SPB infestations was analyzed for each identified cluster.

Unlike other comovement measures, Φ_t constitutes a time series that can be used for further analysis. Besides descriptive statistics of Φ_t, the percentage of comovement during the data period was also computed. Additionally, it was also tested whether the comovement had changed over time. Multinomial logit regression was employed to examine the evolution of SPB infestation comovement over time. To do this, Φ_t was classified into three categories based on its value: positive comovement ($\Phi_t>0$), no comovement ($\Phi_t=0$), and negative comovement ($\Phi_t<0$). Note that relative comovement of SPB infestation risks (Φ_t calculated with normalized SPB infestation risks) could be negative though the original measure of SPB outbreak risks is never negative. Because the primary purpose of this study was to diagnose the possible evolution of the comovement, the following simple multinomial logit model was used

$$\Pr ob\ (\Phi_{it} = j) = \frac{e^{\beta_{0j} + \beta_{1j}t}}{\sum_{k=1}^{3} e^{\beta_{0k} + \beta_{1k}t}},\ j=1,\ 2,\ and\ 3. \quad (3)$$

2.3. Assessing Cyclical Patterns

Spectral analysis, a tool for detecting the existence of periodicity in a time series, was employed to diagnose whether the SPB outbreaks within a spatial unit followed a cyclical pattern. This approach builds on the Fourier analysis, which is concerned with identifying a sum of sine and cosine terms, or the Fourier series representation, to epitomize a series or function (Chatfield 2004). The Fourier transform for the series X_t can be written as

$$X_t = a_0 + \sum_{p=1}^{m}\left[a_p \cos(\omega_p t) + b_p \sin(\omega_p t)\right] \quad (4)$$

where a_o is the mean term, a_p is the cosine coefficients, b_p is the sine coefficients, ω is the Fourier frequencies, and m is the number of frequencies in Fourier decomposition. The amplitude of the pth harmonic is $\sqrt{a_p^2 + b_p^2}$, and the phase of the pth harmonic is $\tan^{-1}(-b_p/a_p)$.

The Fisher Kappa test and the Bartlett Kolmogorov-Smirnov test were performed to test the null hypothesis that SPB infestations in a specific spatial unit are white noise against the alternative hypothesis that the infestations contain a deterministic sinusoidal trend. Rejection of the null hypothesis would suggest statistic evidence of cyclical occurrences of SPB infestations in the spatial unit. For those spatial units that exhibited cyclical outbreak patterns, spectral density functions were derived to depict the periodicities of SPB infestations. All statistical analyses in this study were executed using SAS (SAS Institute Inc. 2007).

3. Results

3.1. Infestation Risk

Table 2 presents the descriptive statistics of SPB infestation risks for each state and cluster (The clustering results will be discussed later). Of the 11 states, Tennessee has the highest annual average infestation rate partially due to large outbreaks in recent years; Alabama and South Carolina rank second and third; and Arkansas has the lowest infestation rate.

Compared to the results derived from earlier data covering the period of 1973-1996 (Gan 2004), the average SPB infestation rate for the entire region (all the 11 states) remains about the same. However, SPB infestation risk in Florida, South Carolina, and Tennessee has went up whereas it has declined slightly in all the other states. A simple regression of SPB infestation risk on time also confirms that the infestation risk in Florida, South Carolina, and Tennessee has increased over time though it has decreased in Alabama. The decline of the SPB infestation rate in the Arkansas-Texas cluster is also evident at the 10% significance level (Table 2).

Table 2. Descriptive statistics and time trends of SPB infestation risks.

State/Cluster	Mean[a] (%)	St dev	Min	Max	Skewness	Kurtosis	Time trend[b] (p-value)
Alabama (AL)	0.158	0.199	0	0.717	1.494	1.316	-0.0080 (0.025)
Arkansas (AR)	0.011	0.020	0	0.097	2.968	10.150	-0.0003 (0.513)
Florida (FL)	0.032	0.093	0	0.505	4.484	22.078	0.0034 (0.044)
Georgia (GA)	0.087	0.120	0.001	0.489	1.815	2.927	-0.0025 (0.257)
Louisiana (LA)	0.087	0.253	0	1.298	4.185	17.863	-0.0025 (0.582)
Mississippi (MS)	0.042	0.058	0	0.239	2.051	3.866	-0.0002 (0.882)
N. Carolina (NC)	0.055	0.083	0	0.396	2.996	9.643	-0.0013 (0.222)
S. Carolina (SC)	0.148	0.304	0	1.786	4.325	21.733	0.0074 (0.066)
Tennessee (TN)	0.468	1.348	0	6.387	3.513	12.790	0.0576 (0.050)
Texas (TX)	0.057	0.112	0	0.602	3.462	13.706	-0.0006 (0.644)
Virginia (VA)	0.049	0.131	0	0.603	3.728	13.495	-0.0007 (0.796)
AL-FL-LA-MS-VA	0.079	0.086	0.003	0.316	1.431	1.209	-0.0021 (0.227)
GA-NC-SC-TN	0.126	0.192	0.008	1.008	3.469	14.824	0.0049 (0.207)
AR-TX	0.039	0.066	0	0.302	2.752	8.597	-0.0024 (0.069)
All 11 states	0.090	0.085	0.011	0.400	1.756	4.381	0.0005 (0.758)

[a]The risk is measured by the ratio (in percent) of the volume killed by SPB to the total pine growing stock volume. The means represent the annual average risk for the data period reported by Pye et al. (2005).
[b]This is the estimated value of β_1 in the simple regression, y (SPB infestation risk) = $\beta_o + \beta_1 t + \varepsilon$.

3.2. Comovement

Table 3 shows the descriptive statistics and percentage of (a) bivariate comovement of SPB infestations between each state and the region's median and (b) bivariate or multivariate comovement within each cluster. Cluster analysis based on McQuitty's similarity and Ward's minimum variance (Figure 3) reveals that SPB infestation comovements in these 11 states can be clustered into three groups: Alabama-Florida-Mississippi-Louisiana-Virginia, Georgia-Carolinas-Tennessee, and Arkansas-Texas.

Table 3. Descriptive statistics of comovements of SPB infestation risks.

State/Cluster	Mean	St dev	Min	Max	Skewness	Kurtosis	Percentage of comovement
Bivariate comovement with the region's median							
Alabama (AL)	-0.107	0.621	-0.736	2.023	1.790	4.094	80.6
Arkansas (AR)	-0.051	0.593	-0.536	1.907	2.140	5.004	67.7
Florida (FL)	-0.098	0.299	-0.357	0.773	1.443	1.642	77.4
Georgia (GA)	-0.091	0.632	-0.690	1.893	1.645	2.690	80.6
Louisiana (LA)	-0.038	0.558	-0.335	2.653	4.151	19.195	67.7
Mississippi (MS)	-0.009	0.744	-0.774	2.391	1.925	3.954	77.4
N. Carolina (NC)	-0.146	0.456	-0.610	1.578	1.782	5.613	87.1
S. Carolina (SC)	-0.087	0.440	-0.487	1.704	2.559	8.692	83.9
Tennessee (TN)	-0.134	0.238	-0.360	0.773	1.755	5.687	61.3
Texas (TX)	-0.154	0.297	-0.518	0.407	0.119	-1.050	67.7
Virginia (VA)	-0.103	0.372	-0.377	1.704	3.984	19.559	64.5

Table 3. Continued

State/Cluster	Mean	St dev	Min	Max	Skewness	Kurtosis	Percentage of comovement
Bivariate or multivariate comovement within each cluster							
AL-FL-LA-MS-VA	-0.088	0.124	-0.327	0.000	-0.951	-0.891	45.2
GA-NC-SC-TN	-0.123	0.148	-0.360	0.090	-0.527	-1.302	54.8
AR-TX	-0.221	0.302	-0.518	0.407	0.475	-1.146	74.2
All 11 states	-0.041	0.095	-0.291	0.000	-2.008	2.301	16.1

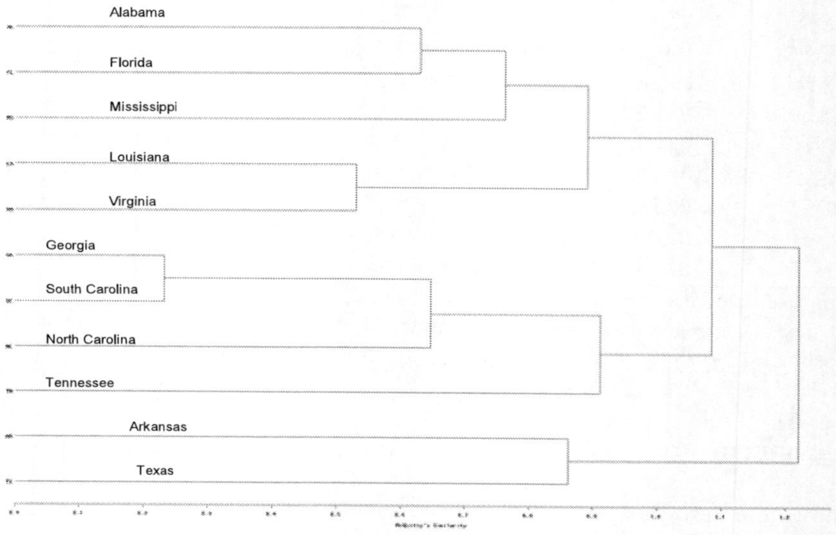

(a) Clustering based on McQuitty's similarity.

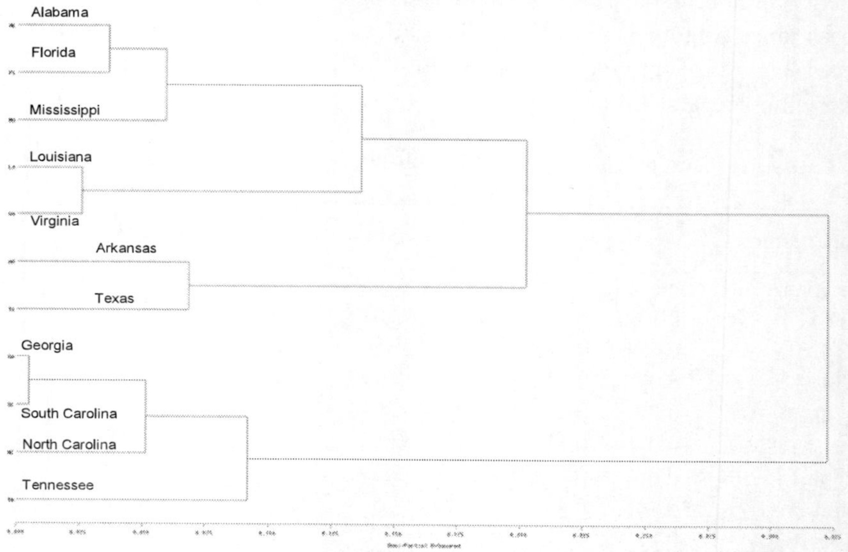

(b) Clustering based on Ward's minimum variance

Figure 3. Clusters of SPB outbreak comovements.

Of the 11 states, North Carolina has the highest percentage of comovement with the region's median, followed by South Carolina, Alabama, and Georgia. Hence, North Carolina can be considered as the region's reference for predicting and monitoring SPB outbreaks in the future. The percentage of comovement with the region's median is lowest for Tennessee, revealing the largest departure of SPB infestations in the state from the regional central tendency. Additionally, the comovements of Louisiana and Virginia with the region's median are highly skewed and leptokurtic. The positive skewness for all bivariate comovements implies that the comovements of large outbreaks between a state and the region's median are more prominent than those of small ones. The opposite is true for the multivariate comovements, suggesting that the comovements of small outbreaks are more pronounced than those of large ones for the entire region and the clusters of Alabama-Florida-Louisiana-Mississippi-Virginia and Georgia-Carolinas-Tennessee (Table 3).

Table 4. Percentage of bi-state comovement of SPB infestation risks.

	AL	AR	FL	GA	LA	MS	NC	SC	TN	TX	VA
AL	100	61.3	71.0	80.6	74.2	77.4	74.2	77.4	54.8	67.7	64.5
AR		100	58.1	61.3	74.2	71.0	61.3	58.1	61.3	74.2	71.0
FL			100	77.4	71.0	74.2	77.4	93.5	77.4	51.6	74.2
GA				100	74.2	77.4	74.2	83.9	67.7	74.2	71.0
LA					100	77.4	54.8	71.0	74.2	61.3	83.9
MS						100	64.5	74.2	58.1	64.5	74.2
NC							100	83.9	67.7	67.7	58.1
SC								100	77.4	58.1	74.2
TN									100	54.8	71.0
TX										100	64.5
VA											100

The percentage of bi-state comovement is presented in Table 4. Overall, the bi-state comovements of SPB infestations in the region are quite strong. SPB infestations in Florida and South Carolina have moved with each other very closely (93.5% of comovement) whereas the percentage of comovement between Florida and Texas is lowest (51.6%). Among the three clusters, the comovement between Arkansas and Texas is strongest whereas the synchronization among Alabama, Florida, Louisiana, Mississippi, and Virginia is weakest (Table 3). There is only about 16% of comovement among all 11 states, suggesting weak evidence of regionwide SPB infestation synchronization.

Multinomial logit regression results indicate that there is no statistical evidence of change in SPB infestation comovement over time for most states and clusters (Table 5). The only exceptions are Florida, Texas, and the Arkansas-Texas cluster. The evolution of comovement over time is found statistically significant at the 5% significance level for Florida and at the 10% level for Texas and the Arkansas-Texas cluster. The comovements between Texas and the region's median and between Arkansas and Texas have lessened over time, whereas those between Florida and the region's median have augmented. Likelihood ratio tests suggest that the models generally fit the data well.

Table 5. Test results for possible change in SPB infestation comovement over time based on multinomial logit regression.

State/Cluster	Chi-square (p-value) (H_0: no change over time)	Likelihood ratio (p-value) (goodness of fit)
Bivariate comovement with the region's median		
Alabama (AL)	2.32 (0.314)	56.39 (0.460)
Arkansas (AR)	1.97 (0.373)	58.41 (0.387)
Florida (FL)	6.00 (0.049)	47.77 (0.775)
Georgia (GA)	0.89 (0.642)	56.10 (0.471)
Louisiana (LA)	0.60 (0.739)	50.57 (0.680)
Mississippi (MS)	1.73 (0.420)	57.14 (0.432)
N. Carolina (NC)	1.94 (0.379)	54.99 (0.513)
S. Carolina (SC)	4.33 (0.115)	47.82 (0.774)
Tennessee (TN)	4.14 (0.126)	41.67 (0.923)
Texas (TX)	4.75 (0.093)	55.57 (0.491)
Virginia (VA)	0.95 (0.621)	51.23 (0.656)
Bivariate or multivariate comovement within each cluster		
AL-FL-LA-MS-VA	0.01 (0.949)	41.05 (0.053)
GA-NC-SC-TN	1.84 (0.399)	43.37 (0.891)
AR-TX	5.17 (0.075)	54.01 (0.551)
All 11 states	0.29 (0.593)	26.74 (0.532)

3.3. Cyclical Patterns

Based on Bartlett's Kolmogorov-Smirnov test results (Table 6), at the 1% significance level the null hypothesis that SPB outbreak risk in a given state follows a random distribution (against the alternative hypothesis that the risk contains a deterministic sinusoidal component) cannot be rejected for 10 out of the 11 states (except North Carolina). At the 5% significance level SPB outbreaks in Alabama and Tennessee also exhibits statistically significant evidence for the existence of a deterministic sinusoidal component. At the 10% level, the null hypothesis is also rejected for Louisiana, Mississippi, and Texas

Table 6. Test results for cyclical patterns of SPB outbreaks.

State/Cluster	Bartlett's Kolmogorov-Smirnov statistic (p-value)	State/Cluster	Bartlett's Kolmogorov-Smirnov statistic (p-value)
Alabama (AL)	0.343 (0.0462)	Tennessee (TN)	0.431 (0.0110)
Arkansas (AR)	0.258 (0.2714)	Texas (TX)	0.280 (0.0635)
Florida (FL)	0.270 (0.1941)	Virginia (VA)	0.131 (0.9602)
Georgia (GA)	0.162 (0.7932)		
Louisiana (LA)	0.322 (0.0721)	AL-FL-LA-MS-VA	0.428 (0.0083)
Mississippi (MS)	0.314 (0.0849)	GA-NC-SC-TN	0.313 (0.1050)
N. Carolina (NC)	0.468 (0.0002)	AR-TX	0.357 (0.0437)
S. Carolina (SC)	0.240 (0.1993)	All 11 states	0.247 (0.3193)

The spectral density functions of SPB infestation risks in those states that demonstrate statistical evidence of cyclical patterns at the 10% significance level or better are presented in Figure 4. The spectral density functions look quite different for many states though most variations of SPB infestation risks are concentrated at low frequencies. On average, Tennessee has the largest amplitude, followed by Louisiana and Alabama whereas Mississippi has the smallest.

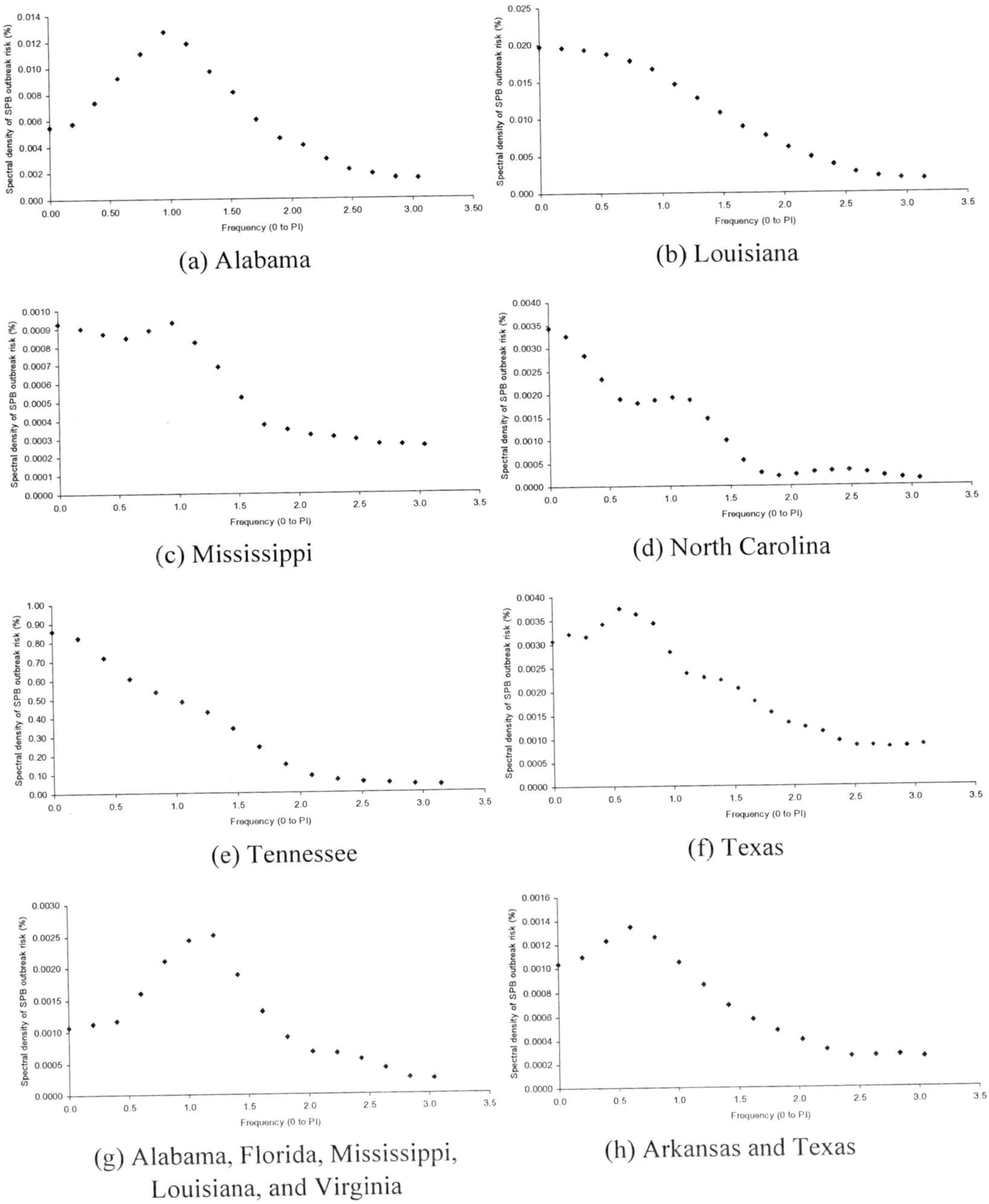

(a) Alabama

(b) Louisiana

(c) Mississippi

(d) North Carolina

(e) Tennessee

(f) Texas

(g) Alabama, Florida, Mississippi, Louisiana, and Virginia

(h) Arkansas and Texas

Figure 4. Spectral density functions of SPB infestation risks.

For Alabama, the spectral density function peaks at a frequency of 1.0, implying that SPB outbreaks in Alabama have been dominated by a deterministic sinusoidal component at a frequency of 1.0 (a period of seven years). The cyclical component over a frequency range of 0.8-1.1 (a period range of six to seven years) accounts for 35% of the total variation. For Louisiana, the spectrum is largest at zero frequency and cuts off at a frequency of 2.5 (a period of about two years). The cyclical component for frequencies lower than 1.0 (periods greater than six years) constitutes almost 70% of the total variation. For Mississippi, the variation of SPB outbreak risk is concentrated at low frequencies and reaches its maximum at a frequency of about 1.0 (a period of seven years). The cyclical component for frequencies lower than 1.5 (periods greater than four years) constitutes 75% of the total variation. Similarly, the spectra of SPB outbreaks for North Carolina, Tennessee, and Texas are also concentrated at low frequencies. For North Carolina, the deterministic sinusoidal component for frequencies lowers than 1.25 (periods greater than five years) accounts for 85% of the total variation. For Tennessee, the deterministic sinusoidal component for frequencies lowers than 1.5 (periods greater than four years) amounts to 90% of the total variation (power). And for Texas, the deterministic sinusoidal component for frequencies lowers than 1.0 (periods greater than six years) constitutes 60% of the total variation.

There is stronger statistical evidence of cyclical patterns in the three clusters than in many individual states. Cyclical patterns are detected in the Alabama-Florida-Louisiana-Mississippi-Virginia cluster at the 1% significance level and in the Arkansas-Texas cluster at the 5% significance level (Table 6). For the Alabama-Florida-Louisiana-Mississippi-Virginia cluster, the spectral density function peaks at a frequency of 1.2 (a period of about five years), with 65% of the variation concentrated in the periods between four and ten years. For the Arkansas-Texas cluster, the density function reaches its maximum at a frequency of 0.6 (a period of ten years), and the sinusoidal component for frequencies lower than 1.1 (periods of greater than six years) makes up two-thirds of the total variation.

4. Concluding Remarks

This study was intended to examine the possible comovement and cyclical patterns of SPB infestations in the Southern United States. A suit of analytical tools were used to detect and characterize these patterns. Here are the key findings. SPB infestation risk remains unchanged over time for the region as a whole thought it has went up for some states and down for some others. Based on the comovements of SPB infestations, the 11 states can be clustered into three groups: Alabama-Florida-Louisiana-Mississippi-Virginia, Georgia-Carolinas-Tennessee, and Arkansas-Texas. The bi-state comovements of SPB infestations and comovements within the three identified cluster are relatively strong though regionwide synchronization is weak. Based on bivariate comovement analysis, North Carolina best resembles the region's median of SPB infestation risks. Furthermore, comovements between Texas and the region's median and between Arkansas and Texas have strengthened over time, whereas synchronization between Florida and the region's median has weakened.

The regionwide SPB outbreaks do not exhibit a cyclical pattern though statistical evidence of cyclical infestations exists in the states of Alabama, Louisiana, Mississippi, North Carolina, Tennessee, and Texas. Even for many of these states that exhibit cyclical outbreak patterns, a single dominant frequency of SPB outbreaks is hard to detect. The variation of

SPB infestation risk spreads over a relatively wide range of frequencies while being concentrated at low frequencies (mostly <1.5). Additionally, cyclical patterns of SPB outbreaks appear more evident in the identified clusters than in individual states or the entire region.

These findings have several important implications for monitoring and mitigating SPB infestations in the region. First, because of its close resemblance with the region's median infestation risk, North Carolina can serve as the region's reference for monitoring and projecting SPB infestations. Second, strong comovement of SPB infestations exists for some subregions in the U.S. South, but not for the entire region as a whole. Hence, in addition to regionwide efforts, different monitoring and mitigation strategies should also be developed and implemented for different subregions or clusters. Third, cyclical outbreak patterns exist only in few states and two clusters and occur over a relatively wide range of low frequencies. This implies future monitoring and mitigation efforts cannot rely on the anticipation of a regular cyclical pattern. It seems obvious that SPB infestations in the region have taken place all the time (at least most time) though the magnitude of infestations varies temporally and spatially. This calls for close monitoring of SPB infestations in the region on a continuing basis.

This study used all the available SPB infestation data for the region. The time span (1960s-2004) of data series might be relatively short for spectral analysis, especially given that the sinusoidal component is concentrated at low frequencies. It is recommended to update these analyses when more data become available in the future. Hopefully, this study has elucidated some important features of spatial and temporal patterns of SPB infestations in the region, which will be of value for guiding future monitoring and mitigation efforts.

References

Baur, D. (2003). What is co-movement? *Working paper*. European Commission, Joint Research Center, Ispra (VA), Italy.

Chatfield, C. (2004). *The Analysis of time series: An introduction* (6th ed.). Boca Raton, FL: Chapman & Hall/CRC.

Dickey, D., and Fuller, W. (1979). Distribution of the estimates for autoregressive time series with a unit root. *J. Amer. Stat. Ass.* **74**, 427-431.

Dickey, D., and Fuller, W. (1981). Likelihood ratio statistics for autoregressive time series with a unit root. *Econometrica* **49**, 1057-1072.

Engle, R., and Granger, C. (1987). Co-integration and error correction: Representation, estimation and testing. *Econometrica* **55**, 251-276.

Everitt, B.S., Landau, S., and Leese, M. (2001). *Cluster analysis* (4th ed.). New York: Oxford University Press.

Gan, J. (2004). Risk and damage of southern pine beetle outbreaks under global climate change. *Forest Ecology and Management* **191**, 61-71.

Gumpertz, M.L., Wu, C., and Pye, J.M. (2000). Logistic regression for southern pine beetle outbreaks with spatial and temporal autocorrelation. *Forest Science* **46**, 95-107.

Mawby, W., and Gold, H. (1984). A reference curve and space-time series analysis of the regional population dynamics of the southern pine beetle (*Dendroctonus frontalis* Zimmermann). *Researches on Population Ecology* **26**, 261-274.

Murray, M.P. (1994). A drunk and her dog: An illustration of cointegration and error correction. *American Statistician* **48**, 37-39.

Price, T., Doggett, C., Pye, J., and Smith, B. (1998). A history of southern pine beetle outbreaks in the southeastern United States. Technical Report, Southern Forest Insect Working Group, Georgia Forestry Commission, Macon, GA.

Pye, J.M. (1993). Regional dynamics of southern pine beetle population. In A.M. Liebhold, and H.R. Barrett (Eds.), *Proceedings of spatial analysis and forest pest management* (pp. 111-124). USDA Forest Service General Technical Report NE-175. Radnor, PA: USDA Forest Service Northeastern Forest Experiment Station.

Pye, J.M., Price, T.S., Clarke, S.R., and Huggett, Jr., R.J. (2005). A history of southern pine beetle outbreaks in the southeastern United States through 2004. Research Triangle, NC: USDA Forest Service Southern Research Station. Available from http://www.srs.fs.usda.gov/econ/data/spb/index.htm.

SAS Institute Inc. (2007). SAS OnlineDoc 9.1.3. SAS Institute Inc., Cary, NC. Available from http://support.sas.com/onlinedoc/913/docMainpage.jsp.

Smith, W.B, Miles, P.D., Vissage, J.S., and Pugh, S.A. (2004). Forest resources of the United States, 2002. USDA Forest Service General Technical Report NC-241. St. Paul, MN: USDA Forest Service North Central Research Station.

USDA Forest Service. (2008). Forest inventory data online (FIDO). Available from http://199.128.173.26/fido/mastf/index.html.

In: Environmental Modelling: New Research
Editor: Paul N. Findley, pp. 103-119

ISBN: 978-1-60692-034-3

Chapter 4

TRENDS IN MODELLING OF RADIONUCLIDES UPTAKE BY PARTICULATE MATTER IN THE MARINE ENVIRONMENT USING BOX MODELS

A. Laissaoui [*] ***and R. El Mrabet***
Centre National de l'Energie, des Sciences et des Techniques Nucléaires B.P. 1382 R.P. 10001 Rabat, Morocco

Abstract

The objective of this paper is to provide a documented discussion on modelling of radionuclides interaction with suspended particulate matter in the marine environment. Aquatic environments occupy a major portion of the earth's surface and, therefore, an understanding of the radionuclide pathways is essential for radioecology purposes. Many radionuclides within such systems exhibit non-conservative behaviour, i.e., they undergo a change in the distribution between the particulate and dissolved phases, due to sorption reactions. The first reactive transport models for non-conservative radionuclides were based upon the equilibrium distribution coefficient, kd. However, recent models are based on kinetic rates of uptake/release of radionuclides between waters and the solid phases when short timescales are involved. Recently, much experimental and modelling effort has been focused on determining those factors which affect the kinetics and the final equilibrium conditions for the uptake of pollutants in aqueous suspensions under dynamic or static situations. Some of the obtained results appear to be either surprising or contradictory and introduce some uncertainty on which parameter values are most appropriate for environmental modelling.

The kinetic box models are widely used in modelling of radionuclides dispersion in aquatic systems, and are based on reversible reactions of constant coefficients. Indeed, several models have been proposed in the literature taking account of parallel, consecutive reactions and some combinations of both depending on the uptake experimental data. Research in this field of radioecology has been stimulated by the high level of public concern over health aspects of the presence of contaminant radionuclides in the environment, which has prompted a generally anti-nuclear posture. In this paper we present a review of the most relevant mathematical models developed so far and a brief revision of the historic developments of theories on surface-electrolyte interactions. Laboratory tracing experiments to study the uptake of ^{239}Pu, ^{241}Am, ^{133}Ba and ^{85}Sr in natural aqueous suspensions from several aquatic

[*] E-mail address: Laissaoui@cnesten.org.ma

systems (reservoir, river, estuary and sea) are also presented to illustrate the application of box models covering a large number of environmental situations and different reaction channels.

Key words: marine radioecology, box models, radionuclides, transfer coefficients.

I. Introduction

Environmental radioactivity studies are of major importance with respect to radioactive waste disposal and future exploitation of nuclear energy. This is due to the high level of public concern over health aspects of the presence of contaminant radionuclides in the environment, which has stimulated a generally anti-nuclear posture. The numerous accidents involving reactors and other nuclear facilities, satellites containing radionuclides, nuclear weapons and large radiation sources (MacKenzie, 2000) have resulted in public distrust regarding the promises of clean energy from the nuclear industry in the mid-20th century.

Aquatic environments occupy a major portion of the earth's surface and, therefore, an understanding of the radionuclide pathways is essential for radioecology purposes. Many radionuclides within such systems exhibit non-conservative behaviour, i.e., they undergo a change in the distribution between the particulate and dissolved phases, due to sorption reactions. Thus, migration of non-conservative radionuclides in aquatic environments is strongly affected by the interaction of the radionuclides with solids present in such systems. Particularly, the uptake and release by particulate suspended mater combined with sedimentation and resuspension of the solids plays a significant role. Indeed, once introduced into the marine environment, particle reactive radionuclides will sorb, more or less strongly, onto solid particles so that they are partitioned between the solid and water. The solid particles can also settle out to the bottom and be removed from the water column. On the other hand, radionuclides in the water can also be adsorbed directly onto the sediment after diffusing across the sediment water interface, always called the boundary layer. Radioactivity can also be released back into the water column by desorption and remobilisation from the sediment and by resuspension of solid material. All of these processes are able to modify the global dynamics of radionuclides and, therefore, must be included in models simulating transport processes. Most of the existing mathematical codes neglect this respect by assuming that the uptake kinetics are so fast that equilibrium is always established.

The first reactive transport models for non-conservative radionuclides were based upon the equilibrium distribution coefficient, kd. This parameter has always been a measure of sorption of contaminants to solids and is defined as the ratio of the quantity of the adsorbate adsorbed per unit mass of solid to the amount of the adsorbate remaining in solution at equilibrium. The widespread use of kd is due to the need for a single parameter describing solid/solution distribution for the purpose of modelling the geochemical partitioning of radionuclides in steady state conditions. However, the use of distribution coefficients is limited by the fact that it can vary by 9 orders of magnitude between different nuclides and 3 orders of magnitude for any particular one (Hamilton-Taylor et al. 1993). This variation depends mainly on the solution composition and the nature of the solid substrate. On the contrary, recent models are based on kinetic rates of uptake/release of radionuclides between waters and the solid phases when short timescales are involved (Perianez, 2004).

Recently, much experimental and modelling effort has been focused on determining those factors which affect the kinetics and the final equilibrium conditions for the uptake of pollutants in aqueous suspensions under dynamic or static situations. Some models include the kinetics as a one-step reversible or irreversible reaction which is often considered as a rough approximation (Laissaoui et al., 1998). However, data reported in the scientific literature show that the interaction of radionuclides with natural solids is often slow and is a two-to three-step process (El Mrabet et al., 2001; Barros & Abril, 2004). Thus, a model taking into account a single reaction is generally unable to properly describe both sorption and release kinetics. It is worth noting that mathematical approaches using box models in the analysis of time-resolved sorption data are capable of providing kinetic parameters which are important in determining the mobility and rate of removal of radionuclides from the water column (Robbins et al., 1992; Smith and Comans, 1996). However, the values of parameters are often different, depending on the number of processes identified in the uptake curves.

In this paper we present a review of the most relevant mathematical models developed so far and a brief revision of the historic developments of theories on surface-electrolyte interactions. Laboratory tracing experiments to study the uptake of ^{239}Pu, ^{241}Am, ^{133}Ba and ^{85}Sr in natural aqueous suspensions from several aquatic systems (reservoir, river, estuary and sea) are also presented to illustrate the application of box models covering a large number of environmental situations and different reaction channels.

II. Theory of Ion Exchange between Water and Suspended Particles

Sorption of radionuclides by particles in aqueous suspensions can be regarded as a physico-chemical process through which the charges on the mineral lattice, generally negative, are compensated by ions of positive charge. This phenomenon can take place on the outer surfaces of the particles as well as on the inner surfaces of the pores, as shown in Figure 1 (Abril ,1998a; Eisma, 1993). Thus, the whole volume of a particle is not really relevant to this process. One of the most common adsorption reactions in aqueous suspensions is ion exchange. In its most general meaning, an ion-exchange reaction involves the replacement of ionic species on a solid phase by another ionic species taken from an aqueous solution in contact with the solid; that is, a previously-adsorbed ion of weaker affinity is exchanged by the particle for a dissolved ion. Most metals in aqueous solution occur as charged ions and thus metal species adsorb primarily in response to electrostatic attraction.

The effect of surface charge on particles is that ions in the aqueous phase distribute themselves so that electrical neutrality is reached. Radionuclide adsorption is a statistical result of ionic collisions over the free surfaces of particles and of the probability of occurrence of different physico-chemical reactions. In most cases, collisions may take place over a non-active site on the surface of the particle and thus the exchangeable ion returns back into solution. Thus, only a fraction of all the collisions will be captured by uncompensated charges (Abril, 1998a).

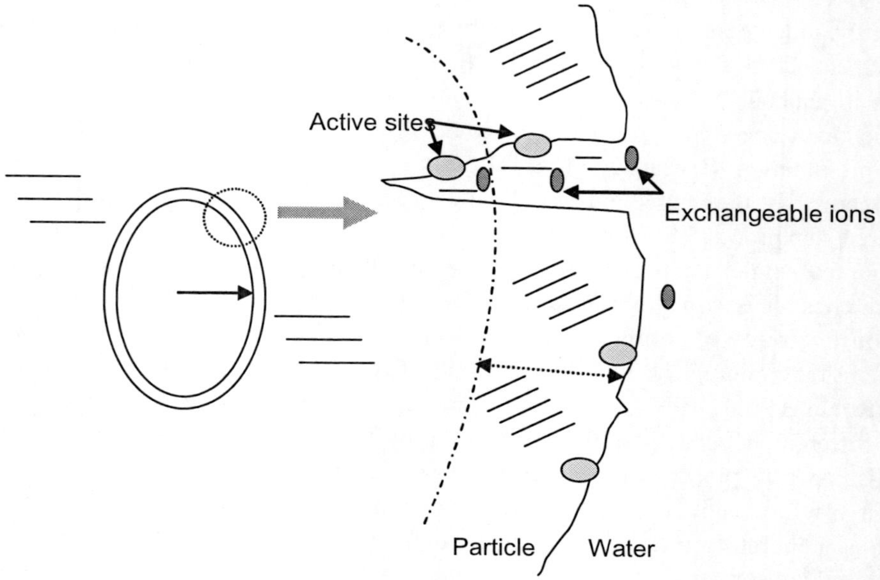

Figure 1. Schematic representation of a particle with its external surface layer.

A revision of the historic development of early theories on surface-electrolyte interactions can be found in Arnold (1978) and Richter (1987). Some of these theories relate only to equilibrium conditions, such as the Gouy-Chapman diffuse double layer and Stern double layer theories or the application of Donan Principles to cation exchange. Non-equilibrium formulations, as presented in Richter (1987), often start from a governing differential equation with constant coefficients, which is stated for each specific adsorption case, such as the Langmuir approach, the Freundlich reaction, for mixed-order reactions, or the competition reactions. Furthermore, of all the complex reactions that take place in heterogeneous systems, the Langmuir adsorption is the most significant. It was originally proposed to describe adsorption of gas molecules onto homogeneous solid surfaces that exhibit one type of adsorption site. Many investigators have tacitly extended the Langmuir adsorption model to describe adsorption of solution species onto solid adsorbents including heterogeneous solids such as suspended particulate matter. It expresses the non-specific adsorption of ions at surfaces; non-specific means that all sorption sites are equal. This feature presents restrictions due to the number of active sites to be occupied which is limited by the available area. For this reason, Langmuir adsorptions show characteristic saturation phenomena. On the other hand, if the adsorption at heterogeneous surfaces is to be described where sites with different adsorption energy exist, then the Freundlich approach will be used to describe the equilibrium state.

III. Development of Kinetic Box Models

Accurate numerical modeling of non-conservative radionuclide behavior in aquatic systems requires the knowledge and quantification of the processes able to modify the global dynamics of such radioactive elements. In this way, some theoretical considerations and experimental studies on the uptake kinetics of dissolved radionuclides have been carried out in recent decades and reported in the scientific literature (Bird & Evenden, 1994; Abril, 1998a and b; Laissaoui et al., 1998; Benes & Cernick, 1992). Therefore, a number of different models have been proposed to qualitatively describe and quantitatively predict the dynamic uptake of radionuclides in aqueous suspensions (Nyffler et al., 1984; Laissaoui & Abril, 1999; El Mrabet et al. 2001; Barros & Abril, 2004). These box models included parallel and/or consecutive as well as reversible uptake kinetic reactions involving one to three steps migration. It may be noted that different models could be fitted to a given experimental data. The goodness of fit of the model should be compared using statistical parameters as the residual sum of squares. This is the simplest criterion for deciding which of the box models is most appropriate (Bunker et al. 2001).

III.1. The One-Step Reversible Reaction

The simplest model (Figure 2) consists of a single reversible reaction quantified by two kinetic transfer coefficients (k_1 for sorption and k_2 for desorption) where the mathematical solution is an exponential decay function for each compartment. The basic assumptions for this case are the following:

- There is no discrimination between reactive sites in the adsorption process.
- There is no interaction between the adsorbed ions.
- In the case of mixed adsorption of different species, they compete for the same sites but competitors for identical sites have different energies.

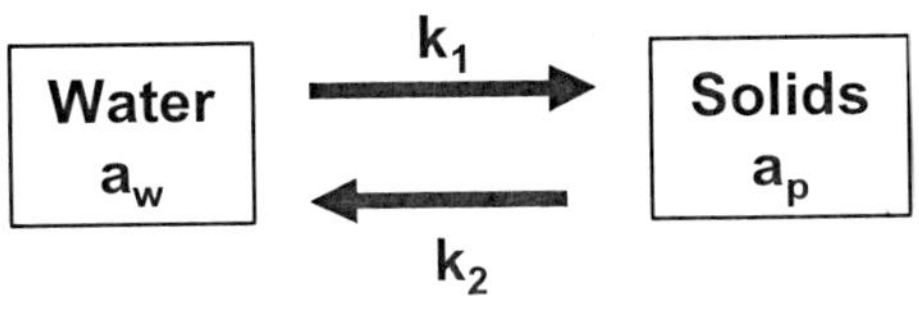

Figure 2. Schematic representation of the one-step model. a_w and a_p are the tracer concentrations in the dissolved phase and in suspended particles respectively. The transfer coefficients k_1 and k_2 are assumed to be constant.

The governing differential equations in the case of a single step model are:

$$\frac{da_w}{dt} = -k_1.a_w + k_2.a_p$$
$$\frac{da_p}{dt} = +k_1.a_w - k_2.a_p \qquad (1)$$

k_1 an k_2 are respectively the forward and backward transfer rates having dimension of $[T^{-1}]$ and can be obtained by fitting the experimentally determined data to the analytical solution of Eqs. 1, which is the following:

$$a_w(t) = \frac{k_1.a_w(0) - k_2.a_p(0)}{k_1 + k_2}\left(e^{-(k_1+k_2).t} - 1\right) + a_w(0) \qquad (2)$$

$a_w(0)$ and $a_p(0)$ are the initial concentrations of the studied radionuclide in the dissolved phase and particles respectively. $a_p(0)$ is always equal to zero since most experimental studies start from uncontaminated particulate matter conditions.

Previous studies devoted to the kinetic aspects of the exchange of radionuclides with particles have been based on mathematical models describing the sorption kinetics using single step reaction model (Jannasch 1988, Laissaoui et al. 1998 among others). This kind of models have shown its applicability in several studies, however, such a mathematical representation fails to describe some experimentally observed phenomena; the sorption of some radionuclides (e.g. Pu and Am) by suspended particulate matter can involve different processes with distinct characteristic time periods: very fast sorption followed by much slower and more extended reaction (Benes et al. 1992). Consequently, to avoid a sub-parameterisation in describing the processes involved in the fixation of radionuclides, it is necessary to consider models of higher orders taking account of the existence of well-defined reaction channels or type of sites on the suspended particles.

III.2. The Two-Step Model

The second model, in order of complexity, is the so called "3-box" or "two-step" model, in which the particle is conceptually divided in two fractions: one associated to the faster reaction channel and another related to the slower uptake. In this case the model, as depicted in Figure 3, can be built in two different ways, considering parallel or consecutive reversible reactions (i.e. four coefficients are needed).

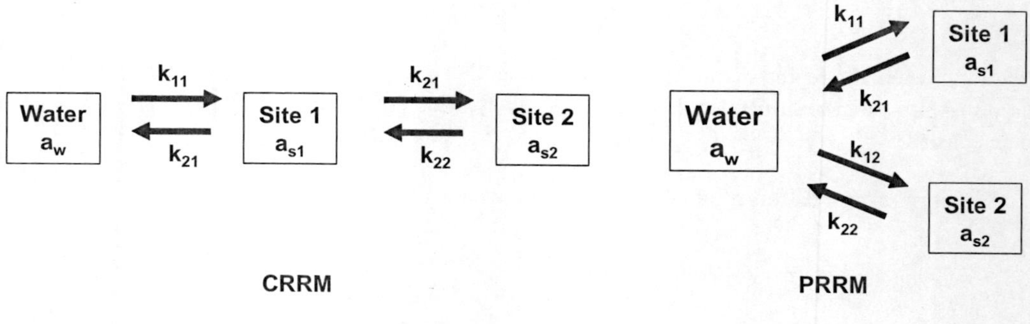

Figure 3. Schematic representation of the two-step models. PRRM is parallel and reversible reaction model and CRRM is consecutive and reversible reaction model.

When taking into account a two consecutive and reversible reaction model (CRRM) for describing interaction of radionuclides at the liquid-solid interface, one supposes the existence of a first mechanism which allows the adsorption of the dissolved ion onto the particles surface. This first reaction is due mainly to electrical forces (such as electrostatic or Van der Waals forces) since dissolved radionuclides occur as charged ions. A second reaction follows which is due to cation exchange or diffusion into the porous structure of the particles. Both reactions are reversible and, therefore, the adsorbed ion can return to the dissolved phase at any step. On the contrary, the two parallel and reversible reaction model (PRRM) assumes that the two reactions involved in the kinetics are independent from each other and are connected only via the cation exchange with the dissolved phase.

The governing equations of both models are as follows:

$$\frac{da_w}{dt} = -k_{11}.a_w + k_{21}.a_{s1} - k_{12}.a_w + k_{22}.a_{s2}$$
$$\frac{da_{s1}}{dt} = k_{11}.a_w - k_{21}.a_{s1}$$
$$\frac{da_{s2}}{dt} = k_{12}.a_w - k_{22}.a_{s2}$$

Eq. (3)
PRRM

$$\frac{da_w}{dt} = -k_{11}.a_w + k_{21}.a_{s1}$$
$$\frac{da_{s1}}{dt} = +k_{11}.a_w - k_{21}.a_{s1} - k_{12}.a_{s1} + k_{22}.a_{s2}$$
$$\frac{da_{s2}}{dt} = +k_{12}.a_{s1} - k_{22}.a_{s2}$$

Eq. (4)
CRRM

Both set of equations are analytically solved by applying Laplace transformation and the solution has the following form:

$$a_w(t) = a_w(t=0).\left[A_1.\left(e^{-P_1.t} - 1\right) + A_2.\left(e^{-P_2.t} - 1\right) + 1\right] \quad (5)$$

P_i and A_i ($i = 1, 2$) are functions of the transfer coefficients k_{ij} and are obtained from the best fit to the experimental uptake curve. The coefficients, in the case of PRRM, can be derived easily from the following set of expressions:

$$\sum_{ij} k_{ij} = -\sum_i P_i$$

$$k_{21}k_{22} + k_{11}k_{22} + k_{12}k_{21} = P_1P_2 \quad (6)$$

$$k_{21} + k_{22} = \sum_i P_i(A_i - 1)$$

$$k_{21}k_{22} = P_1P_2 - \sum A_iP_iP_j$$

Although all the parameters are obtained by processing only the time course of concentrations in the dissolved phase, the evolution of the whole system can be determined by solving numerically the set of Equations (3) or (4). Thus, the curves for the material components can be obtained. It may be noted that expression (5) is the analytical solution of both Equations (3) and Equations (4). Consequently, the PRR and CRR models, which are physically different, are mathematically equivalent.

III.3. The Three-Step Model

The three-step model is the most complex description involving four compartments (the dissolved phase and three sites onto the particles) whose application should be well justified by the experimental data.

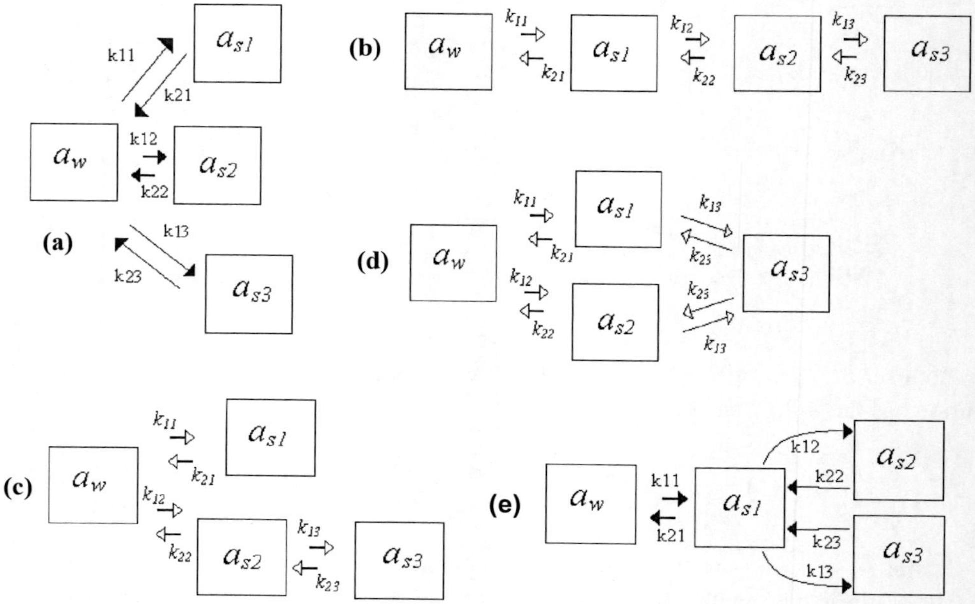

Figure 4. Schematic representation of kinetic box models when four compartments are involved. a_w is the dissolved phase concentration and a_{si} are the concentrations associated to different sites onto the particles.

At least six models could be constructed involving parallel, consecutive reactions and combinations of both of them (Figure 4). It may be noted that there are no selection criteria able to tell about the validity of one model over the others. The set of models are considered as mathematically equivalent since the same multiexponential function could describe the time evolution of concentrations in the dissolved phase independently of the applied model. The question is: what is the physical significance of the kinetic coefficients derived from each model?

A useful illustration of the existence of three-step reaction can be seen in Figure 5 in which the time evolution of ^{133}Ba concentration in the aqueous phase is plotted. The line is the best fit to the sum of three exponential functions reflecting then the superposition of three different adsorption processes in terms of kinetics reaction. The experiment was carried out using natural estuarine waters collected in the Odiel river (southwest of Spain).

Three processes can be clearly distinguished in the uptake curve; a fast adsorption during the first hour followed by a moderated reaction lasting some 20 to 30 hours and finally, an excessively slow reaction which draws the system to a saturation state. The adoption of a model of lower order than the three-step model will be, therefore, a sub-parameterisation of the uptake processes involved in this real case.

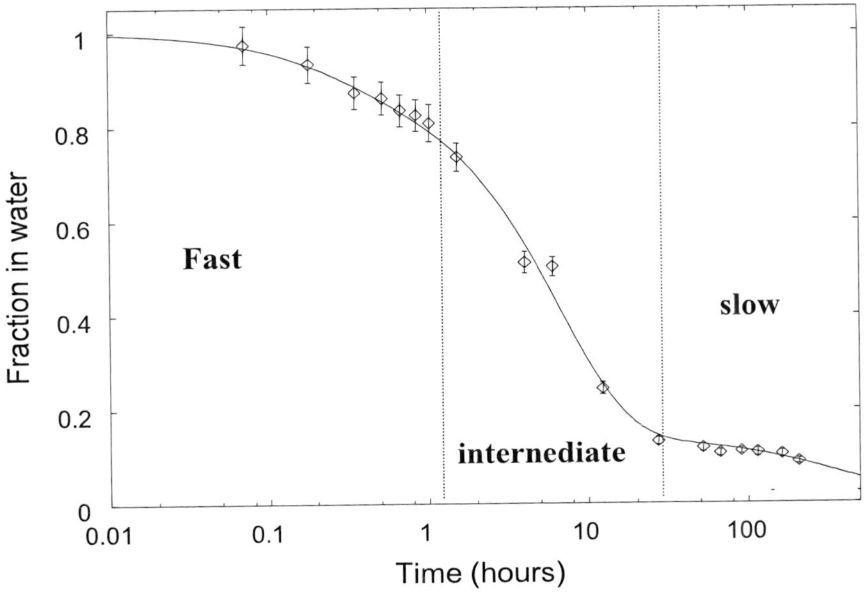

Figure 5. Time evolution of concentrations in the aqueous phase as the result of ^{133}Ba adsorption by suspended particulate matter. The points are the experimental data taken from Barros & Abril, 2004.

Six parameters are obtained from the fitting procedure and are used in the calculation of the kinetic transfer coefficients. By applying the parallel and consecutive models (models a and b in Figure 4), the coefficients k_{ij}(i=1 to 2,j=1 to 3) are obtained and reported in Table 1.

Table 1. Kinetic transfer coefficients k_{ij}(i=1 to 2,j=1 to 3) obtained by applying the three-step parallel and consecutive models to data of Figure 5.

	k_{11}	k_{21}	k_{12}	k_{22}	k_{13}	k_{23}
Parallel (a)	0.365	3.388	0.124	0.019	0.0118	0.0007
Consecutive (b)	0.501	2.472	0.909	0.025	0.0016	0.0007

The computed coefficients were used to solve numerically the governing differential equations in view to determine the evolution of the whole system (the time course of concentrations in the four compartments). Both models are valid in this case and, therefore, the selection of parallel or consecutive description should be arbitrary.

IV. Applicability of Box Models

Kinetically controlled experiments were carried out using radioactive tracers (^{133}Ba, ^{85}Sr, ^{241}Am and ^{239}Pu) to study radionuclides partitioning between natural waters and suspended solid particles. The primordial objective of these studies has been to try to explain the environmental behaviour of radionuclides in natural aquatic environments. Special emphasis has been given to anthropogenic radionuclides which have been present in the environment since 1940s and research into their behaviour, stimulated dominantly by concerns over potential health effects, has been a major area of research in recent decades.

The most commonly used procedure for determining kinetic transfer coefficients is the laboratory batch method using naturally occurring solids and solutions to closely simulate the environmental conditions. It consists of suspending a well characterised sediment of known mass in a known volume of water, or rather using unfiltered water, and adding an identified concentration of an aqueous radionuclide solution. The solution must be continuously stirred with a magnetic stirrer to maintain the particles in suspension. The room temperature should be kept constant during the whole experiment since this physical parameter has a strong effect on the adsorption/desorption. Aliquots of known volume are extracted at controlled elapsed time intervals, transferred to centrifuge vials and immediately centrifuged to separate the solid matter from the liquid. The aqueous phases are finally poured in flasks and the remaining concentration of the radionuclide under study in the supernatant is measured using adequate equipments. It is worth noting that the above procedure provide no certain information regarding the mechanism by which the contaminant are associated to the solid phase.

Strontium

Radiostrontium, particularly its isotope ^{90}Sr, is one of the most intensively studied contaminants in both the terrestrial and marine environments because of its long half life (28.6 years) and its high toxicity (Pröhl et al. 2006, Sysoeva et al. 2005 among others). It is known that it has a relatively weak affinity for solid particles in seawater (Benes & Poliak, 1990) compared to some highly reactive radionuclides such as transuranics. An experiment was carried out by suspending 5.85g of sediment in 1 liter of seawater (the electric conductivity at 20°C was 23 mS) collected from the western Mediterranean Sea. ^{85}Sr was used as a tracer in this experiment (Benkdad et al. 2008).

As can be seen in Figure 6 (a) and (b), the experimental data can be properly fitted to both one step and two step models. The first one (Figure 6a) consists of a single reversible reaction quantified by two kinetic transfer coefficients (k_1 for sorption and k_2 for desorption) where the mathematical solution is an exponential function for each compartment. The second model (Figure 6b) takes into account two reactive sites in the particles (the associated specific activities are a_{s1} and a_{s2} in Bq/kg) which can simultaneously exchange the tracer with the dissolved phase having a specific activity a_w in Bq/l. It may be noted that Sr is a highly hydrated cation at low aqueous concentrations and, therefore, its transfer can be properly described by a simple cation exchange reaction, but this is not always the case. Indeed, in the absence of competitive cations; it has been shown that two reactions of different characteristic times are unambiguously involved in the kinetic evolution of adsorption.

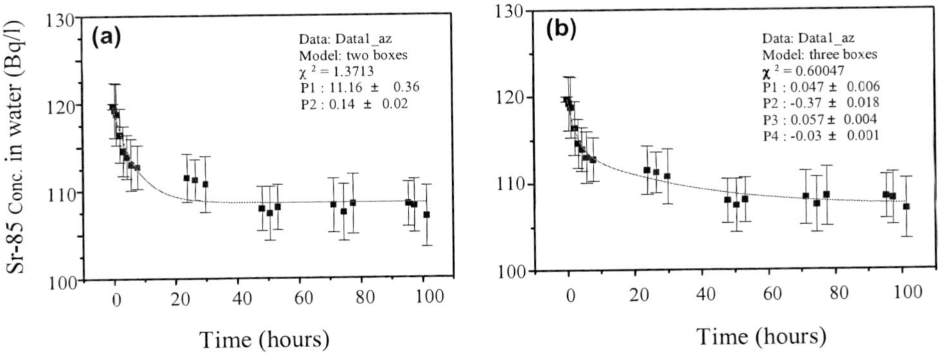

Figure 6. Sorption kinetics of ^{85}Sr in marine aqueous suspension. The points are the experimental data (with 1σ analytical error bars including counting statistics uncertainties) and the lines are the best fit to one-step reversible model (a) and two-step reversible model (PRRM) (b). The P_i values shown in the figure are fitting parameters to Eq. 2 in (a) and to Eq. 5 in (b).

The experimental data (points) were numerically fitted to the theoretical time evolution function (Eq. 2). Table 2 lists the values of the coefficients k_{ij} (in h^{-1}) by using the box model of two-parallel and reversible-reactions. The errors of k_{ij} coefficients were calculated from the uncertainties in fitting parameters P_i (i = 1 to 4).

Table 2. Kinetic transfer coefficients, in h^{-1}, for the uptake of ^{85}Sr in aqueous suspension using naturally occurring solids and solution from the western Mediterranean Sea.

k_{11}	0.0018 ± 0.0003
k_{21}	0.028 ± 0.005
k_{12}	0.0172 ± 0.0015
k_{22}	0.35 ± 0.03
k_d (l/kg)	19.9 ± 2.1

All the parameters were obtained by processing the evolution of ^{85}Sr concentration in the dissolved phase. In view of the major difficulty in measuring experimentally the fraction of pollutant associated with different sites in the particles, the evolution of concentrations in all the compartments (water and the two sites in solids) can be determined by a modeling approach using the computed kinetic coefficients. The set of Equations 3, corresponding to the PRR model, was then numerically solved using a finite differences scheme forwarded in time and considering the initial conditions of our experiment. The time step was 0.01h. Thus, predictions of time evolution of the specific activities in water, the two sites in the particles and distribution coefficient were provided by the model and plotted in Figure 7. From this graph, it can be easily noted a sharp increase of activity in site 2 during the first few hours of the experiment, while adsorption onto site 1 is slower and has an asymptotic shape. This emphasizes the fact that site 2 is more accessible to the tracer than site 1. Thus, site 2 corresponds most probably to the active charges located at the external surface of the particles, and site 1 is related to the micropores in the particles where the tracer is strongly bound.

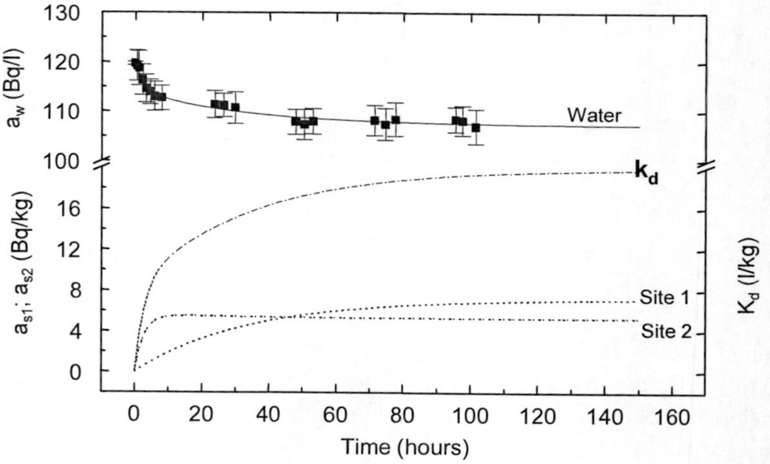

Figure 7. Time evolution of ^{85}Sr concentration in the dissolved phase (line) provided by the two-parallel and reversible-reactions model. The points correspond to the experimental data with 1σ analytical error bars. Predictions of time evolution of specific activities in the two sites and distribution coefficient, kd, are also plotted.

Americium

Americium is an important contributor to radioactivity in nuclear waste with high particle-affinity and complex chemical behaviour in natural systems. The most likely chemical form in the environment is Am(III). Tracer experiments were carried out using unfiltered water samples collected from two different aquatic locations in southern Spain: The Gergal reservoir (NW of Seville – Spain) and the Guadalquivir river (near Seville city).

Figure 8 shows, for both experiments, the time course of the ^{241}Am activity concentration in the dissolved phase. The uptake kinetics of this radionuclide in riverine aqueous suspensions showed three different characteristic times. Consequently, the uptake model must include at least three different reaction channels. In this way, the model (c) of Figure 4 consisting of two parallel and reversible reactions followed by a consecutive adsorption was selected to be applied successfully to the obtained experimental data.

Adsorption is important in the case of the Guadalquivir river experiment because of the high concentration of suspended particulate loads (172 ppm) compared to the experiment corresponding to the Gergal reservoir in which the concentration was only 4 ppm. There is a sharp decrease in the activity of the dissolved phase in the first minutes. In the Gergal experiment the activity in water appears to reach a steady state after few minutes (less than half hour), while in the Guadalquivir experiment a further uptake can be clearly appreciated.

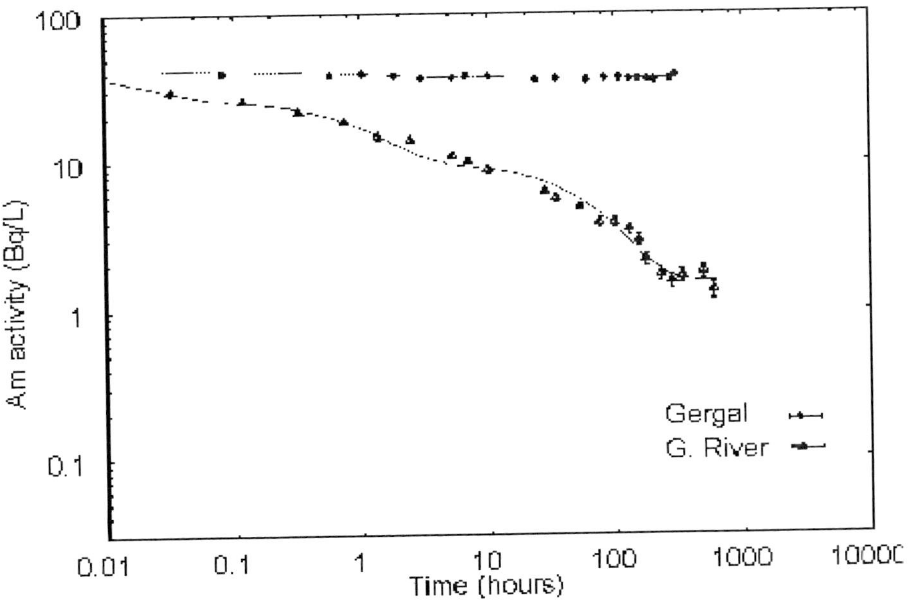

Figure 8. ^{241}Am activity concentrations in the dissolved phase (with 1σ error-bar) corresponding to the tracer experiments using two natural samples (The Gergal reservoir, ms=4 ppm and Guadalquivir river, ms=172 ppm). The lines correspond to the best fits to model (d) (Data from El Mrabet et al. 2004).

Table 3. Parameter values for the model of two parallel reactions plus a consecutive reversible reaction. The coefficients k_{ij} are in h^{-1}. The χ-test values are 0.8 and 1.3 for Gergal reservoir and Guadalquivir river experiments respectively.

	ms (ppm)	$k_{1,1}$	$k_{2,1}$	$k_{1,2}$	$k_{2,2}$	$k_{1,3}$	$k_{3,2}$
Gergal R.	4	0.85 (*)	17.0 ± 1.1	0.07 ± 0.02	0.50 ± 0.15	0.001 (*)	N.I.
G. River	172	40 ± 5	8.5 ± 1.6	1.1± 0.3	0.13 ± 0.05	0.030 ± 0.012	0.0036 ± 0.0022

Values marked with asterisk (*) have errors of the order of 100%. N.I. means that this channel of reaction has not been included in the model.

A two step model, or even a one step reaction model, can describe successfully the experimental data from the Gergal case, while a three-step model is unavoidable to parameterise data from the Guadalquivir experiment. For the sake of homogeneity, the model of two parallel and reversible reaction followed by a consecutive weakly-reversible adsorption (model c in Figure 4) was applied in this study to both experiments for comparison purposes. From Table 3, kinetic coefficients for the forward reaction, $k_{1,j}$ (j=1 to 3), of the first experiment are extremely lower due to the difference in suspended loads concentrations as commented before.

Further experiments, using ^{241}Am as a tracer, were performed to study the effect of suspended load concentration on the adsorption/release rates. The results showed that the kinetic transfer coefficients for sorption increase with the concentration of solids, while desorption coefficients did not show a clear pattern of variation with this parameter.

Plutonium

Plutonium isotopes' behaviour in natural aquatic ecosystems has been widely studied taking account of its oxidation states (Sibley & Sanchez, 1986). At least four oxidation forms Pu(III), Pu(IV), Pu(V) and Pu(VI) exist in aqueous environments, with the ions in hydrated form. The solubility of Pu in natural waters is low and mainly as Pu(V) and Pu(VI). The adsorbed forms are usually in the Pu(III) and Pu(IV) state. Tracer experiments were carried out with ^{239}Pu in unfiltered waters from three aquatic systems (El Mrabet et al. 2001): the Gergal reservoir (NW from Seville), the Guadalquivir river (near the town of Seville) and the Tinto estuary (close to Huelva city). In each experiment, a typical volume of 1 L of water was spiked with a known amount of ^{239}Pu. Results from three experiments are depicted in Figure 9 in logarithmic scale, which shows the time course of the total ^{239}Pu activity, given in counts per minutes (cpm), in the supernatant waters.

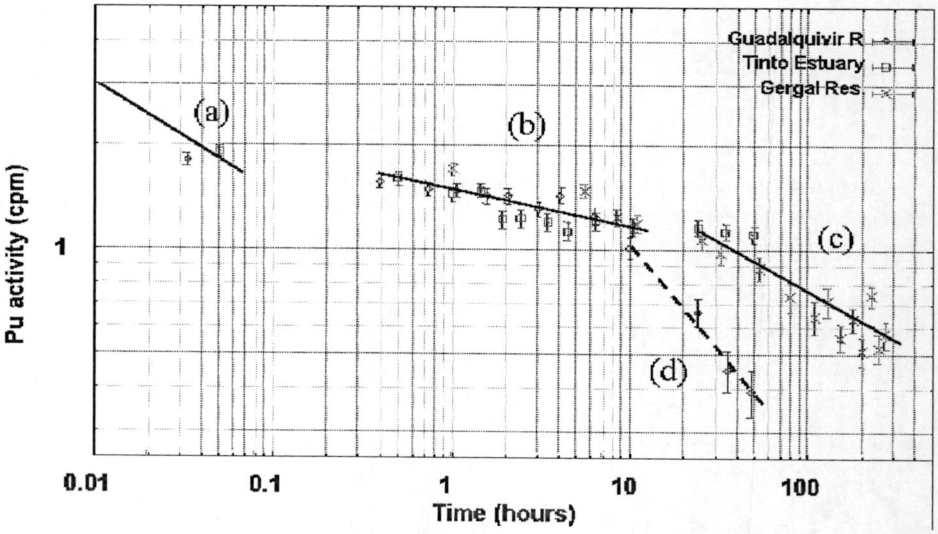

Figure 9. Time evolution of ^{239}Pu activities in the dissolved phase corresponding to the three experiments (points with error bars). Continuous lines indicate the different sequences of the uptake kinetics. Suspended matter concentrations are: 7.37 ppm for the Gergal river, 12.6 ppm for the Guadalquivir river and 7.62 ppm for the Tinto estuary.

The expected initial oxidation state of the tracer is Pu(IV), and an oxidising reaction could follow. The slow increase in concentrations, observed in some cases for long contact times may reflect an oxidation reaction.

Three different regions characterised by different slopes can be distinguished in Figure 9: region (a) corresponds to a fast reaction, lasting for few minutes, during which time up to 30% of dissolved Pu is transferred onto suspended solids; in region (b), a second reaction governs the Pu uptake during 10-20 hours time, transferring up to 30% of remaining Pu in dissolved form. In region (c), the uptake (up to 50% of the remaining Pu in solution) continues during several days. Region (d) corresponds to a faster uptake of Pu from the dissolved phase, which is probably related to flocculation effects. The model of two parallel, reversible and concentration independent reaction followed by a consecutive, non-reversible (with saturation) and concentration independent reaction was successfully applied and the five kinetic coefficients were obtained. In all cases, the first parallel reaction is the faster. It transfers an important fraction of the dissolved Pu to the particulate phase (20% for estuarine waters, 30% for the reservoir and some 42-44% for the river). The period during which this reaction governs the global uptake is about 0.2 h.

The application of a three-step model of single and effective oxidation state in this study allowed reaching a reasonable understanding of the Pu uptake curves. It also provided a set of kinetic parameters useful to describe the most relevant aspects of the electrolytic reactions of ^{239}Pu in the aquatic environment.

V. Conclusion

Kinetic box models are useful approaches to the analysis of time-resolved sorption data and are capable of providing transfer parameters which are important in determining the mobility and rate of removal of radionuclides from the water column. On the other hand, the box models represent an effective way of defining the number of processes involved in the adsorption in aqueous suspensions. A model taking into account a single reaction is generally unable to properly describe both sorption and release kinetics except in some cases such as adsorption of radiostrotium by suspended particulate matter in seawater. Higher order models such as "two-step" and "three-step" are most appropriate since the uptake of radionuclides by solid particles is usually a process where up to three steps can be distinguished: a fast kinetics (minutes), a moderate process (hours) and a slow sorption (days). However, the rationale for the selection of a model instead of another from the same category is still poorly established. This question should be seriously taken into consideration since it is determinant in interpreting the physical meaning of the transfer coefficients. One practical criterion able to discriminate among models is, from the authors' point of view, the statistical analysis of the variability of k_{ij} with physico-chemical parameters. The best model should give a coherent response to changes in these environmental factors.

On the other hand, it may be noted that the use of kinetic transfer coefficients in predictive dispersion models of pollutants has serious limitations which are due to the variability of physico-chemical parameters in natural aquatic systems (conductivity, pH, temperature, etc.) It is well known that rate parameters are sensitive to changes in environmental conditions and, therefore, their inclusion in dispersion models should be in the form of semi-empirical expressions as functions of the most relevant parameters. In this way, many more experimental studies are required to establish quantitatively the variability of transfer coefficients for each radionuclide for a more realistic description of surface-electrolyte interactions.

Finally, investigation on the behaviour of radionuclides in aquatic environments is still of contemporary importance and there will be a continuing requirement for a research capability in this area for as long as nuclear power remains a significant contributor to world energy supply.

References

[1] Abril J.M. (1998a). Basic microscopic theory of the uptake kinetics, transfer and distribution of dissolved radionuclides by suspended matter. Part I: Applications. *Journal of Environmental Radioactivity* **41**, No3, 325-342.

[2] Abril J.M. (1998(b). Basic microscopic theory of the uptake kinetics, transfer and distribution of dissolved radionuclides by suspended matter. Part II: theory development. *Journal of Environmental Radioactivity* **41**, No3, 307-324.

[3] Arnold P.W. (1978). *Surface-electrolyte interactions.* In: The Chemistry of Soil Constituents. (eds.) Greenland and M. H. B. Hayes. Wiley, New York, ISBN 0 471 99619X.

[4] Barros H., Abril J.M. (2005). Constraints in the construction and/or selection of box models for the uptake of radionuclides and metals by suspended particulate matter, *Ecological Modelling*, **185**, 371–385.

[5] Barros H., Laissaoui A. & Abril J.M. (2004). Trends of radionuclide sorption by estuarine sediments. Experimental studies using 133Ba as a tracer. *Science of the Total Environment.* **319**, 253-267.

[6] Barros H. & Abril J.M. (2004). Experimental and modelling study on the uptake and desorption kinetics of 133Ba by suspended estuarine sediments from southern spain. *Water Research* **38**, 749-755.

[7] Benkdad A., Laissaoui A., El Bari H., Benmansour M., IbnMajah M. (2008). Partitioning of radiostrontium in marine aqueous suspensions: laboratory experiments and modeling studies, *Journal of Environmental Radioactivity,* **99**, 748-756.

[8] Benes P. & Cernick M. (1992). Kinetics of radionuclide interaction with suspended solids in modeling the migration of radionuclides in rivers. II: Effect of concentration of solids and temperature. *Journal of radioanalytical and Nuclear Chemistry*, **159** No2, 187-200.

[9] Benes P., Cernik M., & Lam Ramos P. (1992). Factors affecting interaction of radiocesium with freshwater solids. II contact time, concentration of the solid and temperature. *Journal of Radioanalytical and Nuclear Chemistry* **159** (2), 201-218.

[10] Benes P. & Poliak R. (1990). Factors affecting interaction of radiostrontium with river sediments. *Journal of Radioanalytical and Nuclear Chemistry* **141**, 1, 75-90.

[11] Bird G.A. & Evenden W.G. (1994). Effect of sediment type, temperature and colloids on the transfer of radionuclides from water to sediment. *Journal of Environmental Radioactivity* **22**, 219-242.

[12] Bunker D.J., Smith J.T., Livens F.R. and Hilton J. (2001). Kinetics of metal ion sorption on lake sediments – approaches to the analysis of experimental data. *Applied Geochemistry*, 651-658.

[13] Eisma, D. (1993). *Suspended Matter in the Aquatic Environment.* Berlin: Springer.

[14] El Mrabet R., Abril J.M., Manjon G. and Garcia-Tenorio R. (2004). Experimental and modeling study of 241Am uptake by suspended matter in freshwater environment from southern Spain. *Journal of Radioanalytical and Nuclear Chemistry*, **1**, 137-144.

[15] El Mrabet R., Abril J.M., Manjon G. & Garcia-Tenorio R. (2001). Experimental and modelling study of plutonium uptake by suspended matter in aquatic environments from southern Spain. *Water Research* **17**, 4184-4190.

[16] Hamilton-Taylor J., Kelly M., Kershaw P. and Lambert C.E. Radionuclide aquatic pathways. In: *Radioecology after Chernobyl*, (eds.) Sir Frederick Warner and Roy M. Harrisson. (1993), 177-274.

[17] Jannasch H.W., Honeyman B.D., Balistrieri L.S. and Murray J.W. (1988). Kinetics of trace element uptake by marine particles. *Geochimistry and Cosmochimistry Acta* **52**, 567-577.

[18] Laissaoui A. & Abril J.M. (1999). A theoretical technique to predict the distribution of radionuclides bound to particles in surface sediments. *Journal of Environmental radioactivity* **44**, 71-84.

[19] Laissaoui A., Abril J.M., Periañez R., Garcia-Leon M. & Garcia Montaño E. (1998). Kinetic transfer coefficients for radionuclides in estuarine waters: Reference values from 133Ba and effect of salinity and suspended load concentration. *Journal of Radioanalytical and Nuclear Chemistry* **237** 1-2, 55-61.

[20] MacKenzie A.B. (2000). Environmental radioactivity: experiencen from the 20th century – trends and issues for the 21st century. *The Science of the Total Environment* **249**, 313-329.

[21] Nyffler U.P., Li Y.H. & Santschi P.H. (1984). A kinetic approach to describe trace-element distribution between particles and solution in natural aquatic systems. *Geochimica and Cosmochimica Acta* **48**, 1513-1522.

[22] Periáñez. R. On the sensitivity of a marine dispersion model to parameters describing the transfers of radionuclides between the liquid and solid phases. *Journal of Environmental Radioactivity* **73** (2004) 101-115.

[23] Pröhl G., Ehlken S., Fiedler I., Kirchner G., Klemt E. and Zibold G. (2006). Ecological half-lives of 90Sr and 137Cs in terrestrial and aquatic ecosystems. *Journal of Environmental Radioactivity*, **91**, Issues 1-2, 41-72.

[24] Richter J. (1987). *The Soil as a reactor*. Catena Verlag, ISBN 3-923381-09-03.

[25] Robbins J.A., Linder G., Pfeiffer W., Kleiner J., Stabel H.H. and Frenzel P. (1992). Epilimnetic scavenging of Chernobyl radionuclides in Lake Constance. *Geochimistry and Cosmochemistry Acta* **56**, 2339-2361.

[26] Sysoeva A.A, Konopleva I.V. and Sanzharova N.I. (2005). Bioavailability of radiostrontium in soil: Experimental study and modelling. *Journal of Environmental Radioactivity*, **81**, Issues 2-3, 269-282.

[27] Sibley T.H., & Sanchez A.L. (1986). Why do Pu(III+IV) and Pu(V+VI) behave similarly in marine experiments. Letter to the Editor. *Journal of Environmental Radioactivity*, **3**, 75-78.

[28] Smith J.T., and Comans R.N.J. (1996). Modelling the diffusive transport and remobilisation of Cs-137 in sediments: the effects of sorption kinetics and reversibility. *Geochimistry and Cosmochimistry Acta* **60**, 995-1004.

In: Environmental Modelling: New Research
Editor: Paul N. Findley, pp. 121-135
ISBN: 978-1-60692-034-3

Chapter 5

SPATIAL DOWN-SCALING AS A TOOL TO IMPROVE MULTIFUNCTIONALITY INDICATORS IN ECONOMIC MODELS

***Wolfgang Britz*[1] *and Klaus Mittenzwei*[2,*]**

[1] Institute for Food and Resource Economics, University of Bonn, Nussallee 21, DE-53115 Bonn, Germany

[2] Norwegian Agricultural Economics Research Institute (NILF), Postboks 8024 Dep., NO-0030 Oslo, Norway

Abstract

The recent broadening of agricultural policy objectives and their related policy instruments has not been mirrored to the same extent in many agricultural sector models. For example, while the Common Agricultural Policy (CAP) of the European Union is becoming more concerned with rural development (second pillar of the CAP), most agricultural sector models are not able to handle the types of policy instruments covered by the second pillar accurately. Consequently, there is a considerable gap between the demand of policy makers and the supply provided by sector models. This paper aims at contributing to narrow this gap by presenting an extension of the agricultural sector model CAPRI (Common Agricultural Policy Regional Impact Analysis) that enables a deeper regionalization of the model and which may prove useful in extending the relevance and reliability of indicators measuring agriculture's multifunctionality.

Key words: sector modelling, agricultural policy, regional downscaling, multifunctional agriculture, indicators

* E-mail address: klaus.mittenzwei@nilf.no. (Corresponding author)

1. Introduction

The objectives of agricultural policies in Europe have changed over the past decades. While increasing food production and stable farm incomes were important goals in the post-War period, current policy reforms put more focus on the multifunctional aspects of agriculture. For example, the recent reforms of the Common Agricultural Policy (CAP) of the EU claim to promote the so-called multifunctional model of European agriculture. This concept means that agriculture does not only produce food and fibre, but also has a significant impact on inter alia food security, landscapes, the viability of rural areas, the environment and animal welfare (OECD 2001a).

Traditionally, agricultural sector models have proven to be a powerful tool to analyse the economic performance of the farm sector and to study impacts of policy reforms on production, trade and income (van Tongeren et al. 2001). A widely used type are programming models built around the animal and plant activities of primary agriculture and maximising an objective function (e.g. farmer's surplus) under a given set of constraints (e.g. farmland availability, feed and nutrient requirements, policies). The explicit technology description of programming models allows the calculation of different economic and environmental indicators, which can play an important role in quantifying the effects of policy reforms on the performance of the agricultural sector. Agro-environmental indicators have been developed e.g. both by the OECD (2001b) and the EU-commission (2000). The OECD has also proposed indicators for other aspects of multifunctionality (OECD 2001b), while the EU-commission (2006) has recently developed a catalogue of indicators for rural development.

Despite of these demands, the economics literature is sparse on the topic of extending EU-wide agricultural sector models with multifunctionality indicators (Mittenzwei et al. 2007, Britz and Heckelei 2008). The main reason for this shortcoming is twofold. On the one hand, agricultural management decisions and their impacts on the multifunctional aspects show a high variation in space, often interlinked with quite local conditions. For example, environment effects depend heavily on local factors as climate and soil. In addition, the underlying relations are often highly non-linear. This requires a deeply disaggregated regional coverage. On the other hand, analysis of the CAP as a Pan-European policy asks for tools able to simulate impacts for the entire EU-27. Possible approaches are either models covering the EU as a whole or a combination of models for representative sites in combination with an extrapolation algorithm (van Ittersum et al. 2008). In addition, given the share of the EU in global agricultural markets, the integrative modelling of world agricultural markets is necessary in order to capture international price and trade effects of EU policy reforms.

For that reason, the paper discusses the extension of the Pan-European agricultural sector model CAPRI (Common Agricultural Policy Regional Impact Analysis) with regional down-scaling methodologies in order to drive environmental indicator calculators and bio-physical models at an appropriate spatial resolution leading to new and improved multifunctionality indicators.

Several other research projects or approaches deal with linking economic models and their results for larger administrative units to (agricultural) land use at sub-administrative regional scale as e.g. SENSOR (Jansson et al. 2007), EURURALIS (Verburg et al. 2006), INSEA (Adler et al. 2007), GENEDEC (Chakir 2007) or LUMOCAP (van Delden and Luja

2007). However, the combination of Pan-European coverage with the rather disaggregated list of crops, inclusion of stocking densities and output and input coefficient estimates per activity render the CAPRI model rather unique.

This paper proceeds by presenting the CAPRI model, the regional downscaling approach, and the bio-physical model in the next section. The third section provides an overview over the current multifunctionality indicators suggested by the OECD and the European Commission and compares them to those implemented in CAPRI. Section four contains a discussion of the potential impact of the regional downscaling approach on the performance of the current indicators. Finally, the strengths and limitations of the proposed indicators are discussed and areas for future research are addressed.

2. Methods

2.1. The CAPRI Model[1]

The CAPRI model is an agricultural sector model covering the EU-27, Norway and Western Balkans based on non-linear regional programming models consistently linked with a global agricultural trade model. Its principal aim is to analyse impacts of changes in EU (or international) agricultural policies on European agriculture and agricultural markets. Technically, it is a static, partial equilibrium model consisting of four interconnected modules covering (1) regional agricultural supply for EU-27, Norway and Western Balkans, (2) the global and EU markets for major primary and secondary agricultural products including bi-lateral trade, (3) EU markets for young animals and finally (4) CAP pillar I. Henceforth, the modelling system comprised by these four modules will be called the 'core model', while the core model in connection with the regional down-scaling methodologies will be called the 'extended model'.

The supply module consists of non-linear programming (NLP) models comprising about 50 crop and animal activities for each of the around 250 regions on the NUTS II level. The model maximises regional agricultural income at given prices and subsidies, subject to constraints on land, policy variables and feed und plant nutrient requirements in each region. Regional agricultural income is defined as the gross value added (GVA) at producer prices plus direct subsidies (premiums). Costs that are neither included in the GVA nor covered by the restrictions[2] in the NLP models are captured by the introduction of a quadratic costs function, whose slope terms are either estimated from time series analysis (Jansson 2007) or derived from exogenous elasticities. The constant terms are defined such that the model calibrates to a given vector of technical coefficients, levels of the production activities, prices and subsidies.

Direct support and other payments covered by the first pillar of the CAP are calculated for each activity in the policy model. Support measures and other programs that are part of the second pillar of the CAP are currently not covered by CAPRI. In 2007, the first pillar is expected to account for the lion's share of the overall CAP budget with more than 75% of the budget expenditures (EU-commission 2007). However, the share of pillar II measures is to be

[1] This section is based on Mittenzwei *et al.* (2007) and Britz (2005).

[2] Land availability, feed requirements and policy variables (like milk quotas) are modelled as restrictions.

increased considerably by 2013, party due to modulation by which CAP pillar I direct payments are capped and the money saved transferred from pillar I to pillar II.

The global market module covers regionally all countries in world, either as individual countries as e.g. for the single EU Member states, US, Canada, Japan, China, India or as block of countries as e.g. for the African, Caribbean and Pacific (ACP) countries with preferential assess to EU markets. It includes about all primary agricultural products covered endogenously by the supply module plus major secondary products as dairy commodities, vegetables cakes and oils linked to them. The market module is formulated as a spatial multi-commodity model allowing for bilateral trade flows based on the Armington-assumption. It covers major agricultural trade policy instruments such as (bilateral) tariffs, tariff quotas, and subsidised exports. The parameters of the model are collected from other modelling systems, while trend forecasts on prices, yields and other economic variables are based on long-term outlook reports on agricultural markets from the EU-commission (2003) and the Food and Agriculture Organization of the United Nations (FAO 2003).

The third module, the market module for young animals, is based on linking non-linear programming models at EU Member States level derived from the underlying regional models from the supply parts. Subjects to additional constraints linking supply and demand of the different type of young animals exchanged between the animal production activities as piglets or calves, it allows determining market clearing prices for young animals. The policy module, finally, calculates the different CAP pillar I subsidies subject to ceilings in values or physical units derived from the results of the regional programming models.

The overall modelling system is solved by iterating between the different modules. A typical iteration starts with the regional supply models. The results obtained from running the supply module are fed into the market modules by re-calibrating the behavioural functions for agricultural supply and feed demand for the countries covered by regional supply modules. Solving the global market module returns market clearing prices along with supply, demand and trade flows. Finally, premiums are adjusted. The prices and premiums are fed back and drive the next iteration of the supply module. The iterative solution process converges rapidly towards a comparative-static equilibrium solution which is characterised by closed markets and conformity of prices and quantities in both the supply and the market modules.

Major outputs of the global market model include trade flows, market balance positions and producer and consumer prices for the products and countries and country blocks, whereas the supply module delivers for each NUTS II region crop levels and animal numbers with their associated revenues, costs and income at the regional level as well as information about feeding practise and nutrient management.

2.2. Regional Down-Scaling [3]

The down-scaling process disaggregates all major elements of the regional data set from CAPRI to clusters of 1x1 km grid cells for the agricultural area of the EU-27. The process is based on the combination of spatially explicit data as e.g. soil or climatic parameters and either statistically estimated or engineering relations. It ensures that the spatial results are consistently down-scaled from NUTS II level. The approach assumes that market conditions

[3] This section draws heavily on Britz (2008).

and agricultural policies are rather homogenous inside NUTS II regions, so that the spatial distribution of agricultural management characteristics can be derived from local factors as soil type, relief or surrounding land cover. The details of the down-scaling procedures are discussed in Britz (2008) and Leip et al. (2007).

The processing units are sub-regional geo-referenced clusters that are as far as possible homogenous with respect to soil, slope and land cover. Grid cells of the same soil type, slope class, Land Cover class and administrative unit were then combined to pixel clusters building so-called Homogenous Soil Mapping Units (HSMU), covering between from individual 1x1 km cell to several thousand ones. Additional data beyond the delineation features, as climate or soil parameters, were defined per HSMU and stored in data base.

Crop shares are derived by using a two stage statistic estimation framework that allows the breakdown of the CORINE land cover maps to individual crop shares. The first stage derives probability density functions for individual crops as a function of factor like soil and climatic parameters, applying a binary Maximum Likelihood estimator to the LUCAS data. The LUCAS data provided geo-referenced point observations for a large observation sample where single crops had been identified per location based on field visits and photo interpretation. The second step then chooses the most probable combination of crop shares for each HSMU exhausting the given agricultural area at regional level (Kempen et al. 2005).

The crop yield estimation determines the most probable combination of crop yields and irrigation shares per crop and HSMU given potential crop yields from JRC's MARS project for major crops, the irrigation map of the FAO and data on regional irrigation shares. As in the case of crop shares, the sub-regional yields exhaust regional production volumes given the crop shares defined by the step discussed above.

The crop yields are used as explained below as explanatory factor in the estimation of animal stocking densities and drive as well the estimation of crop specific fertilizer application rates. The crop yields at HSMU level are used to derive economic indicators per crop (revenues, variable costs, gross value added, gross value added plus CAP pillar I premiums) based on the assumption of linear input demand functions per crop activity for the different inputs (plant protection, repair costs etc.) and assuming uniform prices for output and inputs inside the regional level. It is planned to add estimates about CAP pillar II payments.

Contrary to the LUCAS sample for crops, no high resolution observation sample for animal stocking densities at Pan-European level is available. In addition, a weak relation between local natural factors as soil and climate and stocking densities can be expected. This is especially true for area independent animal production activities like pigs and poultry. The estimation of stocking densities builds on a cross-sectional regression analysis from the Farm Structure Survey for a mix of NUTS II and NUTS III administrative units with about 500 observations for the EU-27 per animal category. Regression models for the different animal activities in CAPRI as well as aggregates for ruminant and non-ruminants were estimated. The explanatory variables of the regression models include crop and land cover shares, certain crop yields, economic performance indicators for plant production, altitude, slope and climate data, which are offered in different transformation. The estimators were then used in a backward elimination procedure, in which the explanatory variables had been removed as long as the adjusted R squared was increasing or a variable was not significantly different from zero at the 5% level. Own produced fodder and organic fertilizer may be easily transported between the HSMU processing units. The estimation of the local stocking densities are therefore not based on the explanatory variables per HSMU, but on a distance and area weighted average of the area around each pixel cluster. Those locally weighted averages per HMSU were then used to estimate the expected mean and its forecast error for each animal category, providing a priori distribution for the stocking densities per HSMU. A Highest Posterior Density estimator chooses then similar as for crop shares and yields those combinations of stocking densities per HSMU which exhaust the regional herd sizes. The resulting data set was evaluated against out-of-sample from France showing stocking densities for 35.000 single communes based on the FSS. The comparison revealed that the estimation was doing significantly better compared to a solution assigning average regional stocking densities per fodder area for ruminants and average stocking densities per ha for the non-ruminants (Leip et al. 2007). The stocking densities allow to include economic performance indicators for animal activities in the calculation at sub-regional level (revenues, variable costs, GVA and GVA pluss pillar I premiums).

Organic and mineral fertilizer application rates are a highly relevant factor for environmental impacts of agricultural production as they influence realized crop yields and nutrient surpluses, and consequently the whole nutrient and carbon cycle in agriculture. Unfortunately, even at Member State level, data on fertiliser application rates for crops are not available from harmonized European statistics. However, the International Fertilizer Manufacturer Association (IFMA) kindly agreed to let the project team access the results of expert surveys on mineral application rates for crops or group of crops at Member State level. Those data are used in the process of building the regional data base of CAPRI to define regional fertilizer application rates per crop, taking into account regional yields, manure availability, average regional soil parameters and emission factors lined up with the MITERRA and RAINS models (Oenema et al. 2007).

At sub-regional level, the organic and mineral application rates per crop are defined as to recover in average the ones at regional level. Firstly, organic application rates per crop and HSMU are estimated by increasing and decreasing the organic application rate for the crop at regional level depending on two factors. The first factor is the estimated local crop nutrient uptake in relation to the regional one, derived from the crop yield. Crop uptakes are derived from yields. A second factor increase or decreases the rate according to the estimated organic nutrient availability derived from stocking densities and manure excretion coefficients. Here again, as in the case of the estimation of the stocking densities, distance and size weighted averages of the organic nutrient availability around the HSMU are used rather than spot observations. The resulting estimated organic application rates per crop are then scaled in order to recover the given regional rate per crop. In a similar manner, inputs from crop residues, biological fixation and atmospheric deposition are calculated. Finally, the estimated mineral rates are based on the difference between the crop nutrient need and all non mineral sources, corrected by typical loss rates, and a factor based on soil properties. Those estimates per crop are then again scaled to deliver in average the regional mineral application rates.

2.3. Meta-model of DNDC

DNDC (denitrification and decomposition) is a bio-physical model for crop growth with a focus on the nitrate and carbon cycle (Liu et al. 2006). As often with bio-physical models, many processes in DNDC are simulated with a daily or even sub-daily time resolution, and run cover several decades, yielding tremendous processing times for Pan-European applications requiring many site-crop combinations. In order to keep processing time and storage needs in a manageable range, there are two strategies to derive a manageable system layout for large-scale applications of bio-physical models: (1) simulation of selected crops for larger processing units or (2) the development of a statistical response surface from the bio-physical model for the results of interest (e.g., average yearly leaching). An example of the former is "Fate of Pollutants in Terrestrial and Aquatic Ecosystems" (FATE) (Mulligan et al. 2006), where the five dominant crops for 10x10 km grid cells are simulated with the bio-physical model EPIC. Such an approach preserves the full output variables from the bio-physical models, but requires application of the bio-physical model to all crop-site combination again for each scenario to analyze. Due to storage and processor time needs, to handle the number of crop-site combinations is typically limited. A meta-model approach does not only reduce dramatically processing time, but also eases the integration into another IT structure. Many bio-physical models communicate with other applications by input and output files in specific formats, requiring then software to generate and read those files in order to interface with e.g. an economic model. A meta-model however can be easily integrated in the reporting part of an economic model, or if necessary, even linked into its equation system.

There is also a more subtle argument to use a meta-model. In the application presented in the paper, fertilizer rates from the economic model and ex-post consistent to statistical data are used to drive the bio-physical model. Those fertilizer application rates are inter alia derived from the observed yields at NUTS II and, as explained above, from differences in yield potentials at sub-regional scale. Unfortunately, there is no guarantee that the simulated yields from a bio-physical model even in average over all sites in a region would match the regional average. But as the input of nutrients at regional level in the system is fixed via the fertilizer application rates per crop, an error in average simulated crop yields will hence lead to an error in the average estimated nutrient balance, and in nutrient fate (leaching, gas emissions, change in soil stocks). A 10% error in crop yields will already lead to a substantial percentage error in nutrient balance positions.

It is deemed hence important to ensure a match between the simulated yields and the ones used to determine the fertilizer application rates. The meta model offers an elegant way to achieve that by adding both plant specific and other parameters typically used to calibrate the bio-physical model as additional explanatory variable in the regressions. The yield regression equation can now determine those values for the calibration parameters which lead to an exact estimate of the yield at the inputted fertilizer application rates and other farm management parameters. That solution should naturally be combined with a careful calibration of the crop growth model. In that case it has then the potential to reduce possible errors from coupling the different data layers to its minimum.

Regarding the linkage of CAPRI to DNDC, a large observation sample per crop was generated by systematically changing the organic and mineral application rates for typical crop-site combination across Europe. That observation sample was used to estimate per crop independent linear regression models for crop yield expressed in N removal, different gas losses, leaching and the elements of a water balance as percolation and transpiration.

3. Indicators

Indicators have proven useful in evaluating the effects of agricultural policies. They are used to condense, structure and communicate the often complex, highly detailed and intrinsic results of agricultural sector models. Reflecting the policy debate, indicators derived form agricultural sector models focused first on market and income effects. The increased concerns of the environmental effects of agricultural policies led to the development of agro-environmental indicators (EU-commission 2000, OECD 2001a). A broad range of environmental indicators were implemented in CAPRI (Britz 2005). With increased focus on the non-commodity aspects of agriculture, indicators have been developed to cover the broad topics of multifunctionality. The OECD (2001b) lists aspects such as food security, landscape, environment, rural viability, animal welfare, species and ecosystem diversity, soil quality, water quality, air quality, land conservation, cultural heritage and animal welfare. In the EU, rural development indicators have been developed as the policy focus is shifting towards the second pillar of the CAP. The 2006 annual report on rural development in the EU lists 74 indicators on topics like importance of rural areas, socio-economic situation, sectoral economics, environment, diversification and quality of life (EU-commission 2006). Of these 74 indicators, 36 so-called baseline indicators are associated with the four axes of the current rural development policy of the CAP (pillar two). The four axes are (1) improving the competitiveness of the agricultural and forestry sector, (2) improving the environment and the countryside through land management, (3) improving the quality of life in rural areas and encouraging the diversification of economic activity, and (4) LEADER.

The core model of CAPRI contains indicators covering four aspects of multifunctionality: food security, landscape, environment and rural viability (Mittenzwei et al. 2007). In the extended model, existing indicators are at the one hand now available at a higher spatial resolution (HSMU level), while new indicators have been added as well. Those refer specifically to the results obtained from DNDC regarding the Nitrogen and water cycle. Further on, an index relating to the characterization of High-Nature Value Farmland is currently under development which inter alia is based on a Shannon index for the diversity of the annual crops.

The set of indicators at HSMU level is almost identical with those implemented at regional level. Based on DNDC meta models, a detailed break-down of gaseous Nitrogen losses could be added, as well as indicators relating to the water cycle. The HNV indicators with its different sub-indicators is still in a prototype phase. A Delphi approach involving experts from the bio-diversity field is in many instances needed to properly define thresholds and aggregation of the sub-indicators.

Table 1 presents the indicators currently implemented in CAPRI.

The indicators of the core model are described in detail in Mittenzwei et al. (2007). Therefore this paper focuses on the additional indicators derived from the extended model. Those data can be exploited e.g. as interactive maps.

Table 1. List of indicators in CAPRI

Topic	Indicator	Definition	Modell
Food security	Food self-sufficiency	Consumption of domestic produced food in % of total domestic use, based on the energy (kcal) content	Core
Food security	Arable land	ha per 1000 capita	Core
	Grazing animals	LU1 per 1000 capita	
Landscape	Shannon diversity index (H)	$H = -\sum p_i \ln p_i$	Core, Extended
	Grazing intensity	LU[1] grazing animals per ha grass land	Core Extended
	Nature Value Farmland characterization	Intensity index	Extended
	Livestock densities	LU per ha UAA (total, ruminants, non-ruminants)	Core, Extended
	Crop shares	% (single crops, crop aggregates)	Core, Extended
Rural viability	Agricultural revenues	€ per ha	Core Extended
	Agricultural labour force	Annual Working Units	Core
	Intermediate input cost	€ per ha	Core Extended
	Gross value added plus CAP pillar I premiums	€ per ha	Core Extended
Environment	Mineral fertilizer consumption	kg per ha (single crop, UAA)	Core, Extended
	Irrigation water use	ltr per ha UAA	
	Nitrogen balance	kg N per ha UAA	
	Phosphor balance	kg P per ha UAA	
	Ammonia Emissions	kg NH4 per ha UAA	
	Global Warming Potential (GWP)	GWP per ha UAA	
	Change in soil content	kg N per ha UAA	
	Emission of N2O, NO, and N2	kg N per ha UAA	
	Rainfall and Irrigation, leaching, transpiration, evaporation	ltr per ha UAA	

1 Livestock Units (European Commission 1995).

2 p_i is the percentage share of crop *i* out of the total agricultural area.

3 Global Warming Potential based on GHG emissions in CO2 equivalents

4 based on DNDC meta- model

There is a growing concern about loosing the so-called High Nature Value (HNV) farmland (Hoogeveen et al. 2004). HNV in combination with biodiversity is one of the priorities in axis 2 of the rural development programs until 2013. JRC and the European Environmental Agency (EEA) have produced in the period 2005-2007 a Pan-European HNV farmland map on the basis of a regional stratification of CORINE land cover data, biodiversity data (NATURA 2000, Important Bird Areas, Prime Butterfly Area) and national data sets where available (Paracchini et al. 2006). A complementary approach discussed here is based on deriving an index characterizing the probability of HNV presence. The index is an area weighted average of sub-indices for arable crops, fodder areas, olive groves and other permanent crops. The sub-index for arable crops is based on the average application rates of mineral and organic nitrogen per ha and a Shannon index characterizing simultaneously the number of crops grown and their distribution. The index for fodder areas uses a potential yield estimate for fodder under rainfed agriculture in combination with stocking densities to characterize the intensity of fodder production. For olive groves, remote sensing based characterization of the intensity of cultivation could be used (Weinsteiner et al. 2007), whereas for the rest of permanent cultures, again mineral and organic nitrogen input characterize the intensity. The different elements were all mapped into 0-10 indexes, and then aggregated into the final index.

4. Indicator Performance

Many of the indicators calculated in the extended model using regional-downscaling and the connection to bio-physical models, are also available in the core model of CAPRI at the NUTS II level. Indicators related to landscape issues and the environment, depend quite often on local factors. It is there reasonable to believe, that the performance and reliability of those indicators is improved when calculated in the extended model. The nutrient surpluses in Germany illustrate this point. The surplus of Nitrogen per ha is calculated between around 55 and 175 kg at the level of the 35 NUTS II regions (map 1). The regions with the highest nitrogen surplus can be found in the Northern part of Germany, the North-Western part and in Southern Germany.

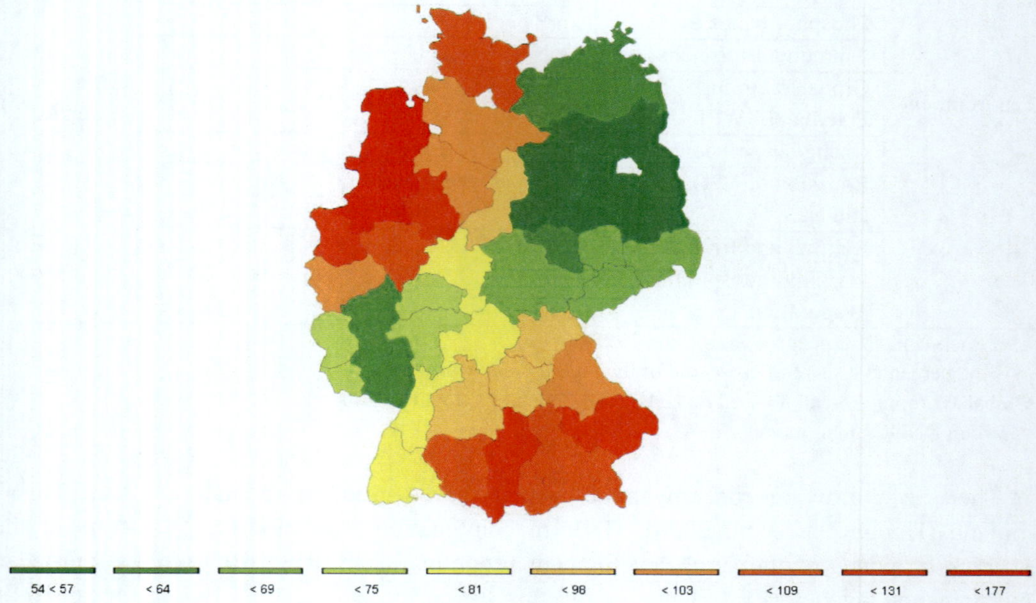

Map 1. Nitrogen surplus in Germany at NUTS II level, average 2001-2003

Map 2 indicates that the numbers reach from slightly below zero about to almost 700 kg at the around 12.000 regions at HSMU level for Germany, of which some 8.300 comprise agricultural land. It should be noted that the colours in map 1 and map 2 depict different levels of nitrogen surplus. In map 1, the dark red area indicates the highest nitrogen surplus (130 – 175 kg N pr ha). In map 2, the same colour depicts values of nitrogen surplus between 290 and almost 680 kg N pr ha. It is therefore necessary to compare the relative range of the regions in the two maps.

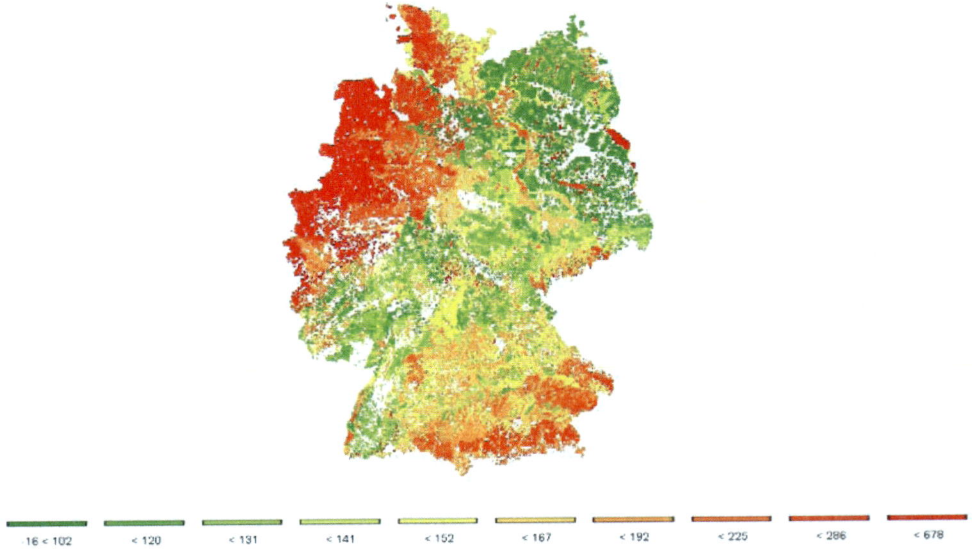

Map 2. Nitrogen surplus at HSMU level, Germany, average 2001-2003.

Comparing the two maps shows that averaging over larger regions levels out local variance and hides the hot spots. For instance, Eastern Germany ranges among the lowest regions at the NUTS II level (green area), but has some extreme high Nitrogen surplus spots in it at the HSMU level at the Eastern part and in Southern-Eastern Germany (dark red spots). Similarly, map 2 shows significant differences for Northern Germany consisting of spots below 140 kg N and spots up to 680 kg N. At the NUTS II level, the same region is just somewhat above national average. For the Central part of Germany, the different regional levels are less pronounced. This is also partly true for the Southern part of Germany, where many regions keep their relative range. That means that below (above) national average Nitrogen surplus areas at the NUTS II level are still below (above) national average at the HSMU level. A possible reason for the significant shifts of the regions in Northern Germany compared to the less pronounced shifts in Central and Southern Germany, is the predominance of area independent pork and poultry production in Northern Germany. In some cases, the NUTS II regions also comprise a mix of rather extensive grass land and intensive crop-livestock production systems.

As the regional HSMU are clustered around soil, slope and land cover, it can be expected that indicators that are highly correlated to these measures will exhibit best performance. This may in particular be true for environmental indicators such as nutrient balances, global warming potential, and water balances.

Indicators on other aspects of multifunctionality, like food security and the viability of rural areas, are probably less affected by the regional down-scaling approach. For the purpose of food security, the regional dimension is most often not very relevant. Only with respect to natural catastrophes and isolation, regional food security may be of importance. In case of the viability of rural areas, the HSMU regional units do not take into account the characteristics of socio-economic variables like the rural economy and the rural labour market. Therefore, the HSMU may be very heterogenous with respect to food security and the viability of rural areas. In conclusion, indicators on these two aspects of multifunctionality at the HSMU regional level may not be very interesting and reliable.

5. Technical Solution

The different down-scaling procedures are realised in GAMS (General Algebraic Modeling System), a modelling language widely used by economic modellers, in which also the CAPRI model is implemented. One major advantage of GAMS is the transparent link to non-linear Programming solvers which are needed for the Highest Posterior Density estimator maximising the joint probability density under a set of constraints. For larger regions with a high spatial variation of the delineation factors, the resulting optimisation problems can be rather large including several ten thousand variables. The whole down-scaling process for one regional data set requires about 18 hours on a high-end work station. The programs are set up so that they can digest ex-post data, projection and scenario results from the economic model.

6. Conclusion

The down-scaling approach of the regional results of the core model of CAPRI together with the link to bio-physical models has improved the overall CAPRI modelling system in two ways. Firstly, the relevance and reliability of many landscape indicators and environmental indicators of the model have been improved by a deeper disaggregated regional level. That allows e.g. to analyse nutrient surpluses in a spatial context taking vulnerable zones into account. The provision at input data at a higher spatial resolution eases linkage to bio-physical models which carry the potential to improve the indicator calculators used at regional level. Secondly, it has been possible to include new landscape indicators and environmental indicators that are highly local in nature.

There are major differences compared to other existing approaches to downscale agricultural management from administrative regions to smaller units: (1) the combination of large-scale application at the level of EU27 with a 1x1 km resolution, (2) the coverage of a rather detailed lists of crops, (3) the inclusion of further management parameters as animal stockings densities and fertiliser application rates, and (4) the application of advanced statistical methods to derive a consistent downscaling results. In opposite to many other approaches land use in CAPRI is depicted by shares, deemed the most suitable presentation when arable cropping is involved. As mentioned above, validation of both the cropping share and the stocking density proved that the connected results are reliable in average.

Unfortunately, evaluation of two further key parameters – the crop yields including grassland yields and fertilizer application per crop – was up to now not possible as high resolution data were not available.

These results indicate that the combination of the core CAPRI model with bio-physical models is worthwhile in improving indicators for selected aspects of multifunctionality. Therefore, the indicators covering other aspects of multifunctionality could potentially be improved using a similar approach.

With respect to the viability of rural areas, more information about the general rural economy including the regional labour market would be needed. This information could be derived, for example, by linking regional CGE (Computable General Equilibrium) models, in which results from CAPRI could be used to feed regional CGEs in order to evaluate the effects of changes in agriculture on the rural economy. This would probably require different regional units than the ones used for landscape purposes and environmental concerns. Food security is typically a national aspect with little local impact. Only for the case of catastrophes that lead to lasting regional isolation, food security is a topic. Given today's infrastructure securing the transport of food over long distances, the possibility of such a case seems quite low.

A further important question is the aggregation of indicators in space. Only in the case of a global externality as gas emissions contributing to global warming, physical quantities may be simply added up to higher regional level. For most other externalities however, aggregation is far more complex as local conditions are needed to be taken into account.

Acknowledgements

The result and methodology presented in here are the outcome of the EU co-funded research project CAPRI-Dynaspat (FP VI 501981). The definition of the processing unit and the necessary geo-processing was handled to a large extent by Adrian Leip, Renate Koebl and Giulio Macchi, all at the JRC. Markus Kempen (University Bonn) was the main responsible for developing the cropping share estimator. Declan Mulligan (JRC) and Adrian Leip (JRC) developed the link to DNDC and provided the input for the meta-model estimation. Maria Luisa Paracchini (JRC) contributed to the development of the HNV indicator.

Errors and omissions remain the sole responsibility of the authors.

References

Adler et al. (2007), INSEA (Integrated Sink Enhancement Assessment), Final Report, IASSA

Britz, W. (2008). EU-wide spatial down-scaling of results of regional economic models to analyze environmental impacts. Paper prepared for presentation at the 107th EAAE Seminar "Modeling of Agricultural and Rural Development Policies". Sevilla, Spain, January 29th -February 1st, 2008.

Britz, W. (Editor) (2005). CAPRI Modelling System Documentation. (http://www.agp.uni-bonn.de/agpo/rsrch/capri/capri-documentation.pdf, as of August 31, 2006).

Britz, W. & Heckelei, T. (2008). Recent Developments in EU Policies – Challenges for Partial Equilibrium Models. Paper prepared for presentation at the 107th EAAE Seminar

"Modeling of Agricultural and Rural Development Policies". Sevilla, Spain, January 29th -February 1st, 2008.

Chakir R. (2007): Spatial downscaling of Agricultural Land Use Data: An econometric approach using cross-entropy method. Toulouse: Inra.

EU-commission (2007). Agriculture in the European Union. *Statistical and Economic Information* (2007). Brussels.

EU-commission (2006). Rural Development in the European Union - *Statistical and Economic Information* - Report 2006. Brussels.

EU-commission (2003). Prospects for agricultural markets 2003-2010. (http://europa.eu.int/comm/agriculture/publi/archive/index_en.htm, as of January 9, 2004).

EU-commission (2000). Indicators for the Integration of Environmental Concerns into the Common Agricultural Policy. *Communication from the Commission to the Council and the European Parliament.* COM(2000)20 final. Brussels, 26.01.2000.

FAO (2003). *World agriculture: towards 2015/2030. Summary report.* Food and Agriculture Organization of the United Nations. Rome.

Hoogeveen, Y., Petersen, J.E., Balazs, K. & Higuero I. (2004*). High nature value farmland - Characteristics, trends and policy challenges*. EAA-Report 1/2004.

Jansson, T. (2007). *Econometric specification of constrained optimization models.* Dissertation. Institute for Food and Resource Economics, University of Bonn, Bonn.

Jansson, T., Bakker, M., Hasler, B., Helming, J., Kaae, B., Neye, S., Ortiz, R., Sick Nielsen, T., Verhoog, D. & Verkerk H. (2007). Description of the modelling chain. SENSOR Deliverable 2.2.1. In: Helming K , Wiggering H, (eds.): SENSOR Report Series 2006/5, http://zalf.de/home_ip-sensor/products/sensor_report_series.htm, ZALF, Germany

Kempen, M., Heckelei, T. & Britz, W. (2005). A Statistical Approach for Spatial Disaggregation of Crop Production in the EU. In Arfini F. (ed), Modelling Aricultural Policies: State of the Art and New Challenges". Parma: Monte Universita: 810-831

Leip, G., Marchi, R., Koeble1, M., Kempen, W., Britz, W. & Li, C. (2007). Linking an economic model for European agriculture with a mechanistic model to estimate nitrogen losses from cropland soil in Europe. In: Freibauer, A., Valentini, R., Dolman, H. and Janssen I. (eds). Greenhouse gases in the northern hemisphere. *Biogeosciences.* Special Issue

Liu, Y., Yu, Z., Chen, J., Zhang, F., Doluschitz, R. & Axmacher, J. C. (2006). Changes of soil organic carbon in an intensively cultivated agricultural region: A denitrification-decomposition (DNDC) modelling approach. *Sci. Total Environ.*, **372**, 203–214

Mittenzwei, K. Fjellstad, W., Dramstad, W., Flaten, O., Gjertsen, A.K., Loureiro, M. & Prestegard, S.S. (2007). Opportunities and limitations in assessing the multifunctionality of agriculture within the CAPRI model. *Ecological Indicators* **7** (2007), 827-838

Mulligan, M., Bouraoui, F., Grizzetti, B., Aloe A. & Dusart, J. (2006). *An Atlas of Pan-European Data for Investigating the Fate of Agrochemicals in Terrestrial Ecosystems*, ILuxembourg: Office for Official Publications of the European Communities, EUR 22334 EN

OECD (2001a). *Multifunctionality. Towards an analytical framework.* Organization for Economic Co-operation and Development, Paris.

OECD (2001b). Environmental Indicators for agriculture – Volume 3: Methods and results (The Stocktaking Report), Organization for Economic Co-operation and Development, Paris.

Oenema, O., Oudendag, D.A., Witzke, H.P., Monteny, G.J., Velthof, G.L., Pietrzak, S., Pinto, M., Britz, W., Schwaiger, E. , Erisman, J.W., de Vries, W. , van Grinsven J.J.M. & Sutton M. (2007). *Integrated measures in agriculture to reduce ammonia emissions.* Final summary report. Wageningen: Alterra

Paracchini, M.L., Terres, J.M., Petersen, J.E. & Hoogeveen, Y., (2006). Background document on the methodology for mapping High Nature Value farmland in EU27. European Commission Directorate General Joint Research Centre and the European Environment Agency.

Van Delden, H. & Luja P. (2007). Integration of multi-scale dynamic spatial models for land use change analysis and assessment of land degration and socio-eonomic processes. In: Proceeding from the conference on Soil protection strategy - needs and approaches for policy support, Polawy, Poland 9-11th March 2006.

van Ittersum, M.K., F. Ewert, T. Heckelei, J. Wery, J. Alkan Olsson, E. Andersen, I. Bezlepinka, F. Brouwer, M. Donatelli, G. Flichman, L. Olsson, A.E. Rizzoli, T. van der Wal, J.E. Wien & Wolf, J. (2008). Integrated assessement of agricultural system – A component-based framework for the European Union (SEAMLESS), *Agricultural Systesm* **96**: 150-165.

van Tongeren, F., van Meijl, H. & Surry, Y. (2001). Global models applied to agricultural and trade policies: a review and assessment. *Agricultural Economics* **26**, 149-172.

Verburg, P.H., Schulp, C.J.E., Witte, N. & Veldkamp A (2006). Downscaling of land use change scenarios to assess the dynamics of European landscapes. *Agriculture, Ecosystems & Environment*: Volume 114, Issue 1, 39-56

Weissteiner, C. J., Sommer, S. & Strobl, P. (2007). Time series analysis of NOAA AVHRR derived vegetation cover as a means to extract proportions of permanent and seasonal components at pixel level. *EARSeL eProceedings*, in press.

In: Environmental Modelling: New Research
Editor: Paul N. Findley, pp. 137-149
ISBN: 978-1-60692-034-3

Chapter 6

ADAPTIVE CONTROL METHODOLOGY AND SOME APPLICATIONS IN ENVIRONMENTAL MODELLING

P.L. Kunsch*
Vrije Universiteit Brussel (VUB) – MOSI Dept. Pleinlaan 2,
BE-1050 BRUSSELS (Belgium)

Abstract

This article summarises the main author's contributions in the past years over environmental modelling within the framework of the methodology called ACM ('Adaptive Control Methodology'). ACM combines the instruments of System Dynamics, multi-criteria analysis and control. Ex-ante control, i.e., prior to policy implementing is called 'control by structure' because it is achieved by developing control feed-back loops in the systemic structure under focus to influence the system evolution towards sustainable long-term paths. Ex-post control takes place during the post-implementation phase in which adjustments of the policy are necessary. ACM contributions in environmental modelling are briefly introduced, and a practical environmental case is presented on the way this technique can be used in real situations.

Keywords: Environmental modelling, Adaptive control, System Dynamics, Multi-criteria Decision Aid

1. Introduction

In ref. [1], the authors plead for developing within Operations Research (OR) new policy instruments able to cope with the challenges of sustainable development in the 21st century. Adequate policies must equally protect the interests of present and future generations and, therefore, they may be called 'ethical'. The authors of [1] further argue that the adequate OR techniques must combine systemic techniques, like provided by System Dynamics (SD) (ref. [2], [3]), with Multi-criteria Decision Aid (MCDA) (a review of common techniques is

* E-mail address: pkunsch@vub.ac.be

provided in ref. [4]), in order to take into account conflicting stakeholders' interests. Also decisions should not be one-shot but continue after a policy has been implemented with monitoring of the policy results and adjusting it when required by the observed system evolution. The methodological framework called Adaptive Control Methodology (ACM) developed in ref. [5] to [7] is proposed to fulfil these requirements. In section 2 the main ACM features are presented. Several contributions of the author in environmental modelling within this framework (ref. [8] to [12]) are briefly reviewed. In section 3 a further case study in energy system modelling is presented. In section 4 conclusions and future prospects are provided.

2. Adaptive Control Methodology

2.1. The Methodology

Detailed presentations on ACM are given in ref. [5] to [7]. Ref. [7] presents in addition a detailed full application to the CO_2 emissions and fossil-fuel usage in the residential sector. The ACM has only been partially applied in the environmental-modelling applications of ref. [8] to [12]. An overview of all applications is given in sub-section 2.2. The main steps of the on-line control are shown in the simplified flowchart of Fig. 1. ACM is an adaptive approach in the sense that at each step in the flowchart backtracking to any upstream step is possible. This occurs when improvements in modelling, strategy revisions, or on-line control operations are required. The full three-stage sequence is the following:

Stage I: elaboration and dynamic modelling;
Stage II: generation and selection of policies by simulation and MCDA;
Stage III: implementation, monitoring and on-line control.

The first stage (I) of system modelling starts with the identification and boundary definition of the analysed problem. It is mainly a development phase producing a mental model shared by analysts and decision-makers. It emphasises anticipation, vision of the future, and coherence. The shared mental model is developed to a working vehicle, using SD simulation software like [3].

Feedback loops are evidenced and analysed to provide insight into the dynamic hypothesis [2]. This includes the analysis of dominant virtuous or vicious circles, the definition of decision variables, and the identification of critical 'watchdogs', i.e., critical indicators delivering warning signals by threats on the viability, or sustainability. This extensive modelling process provides a first SD model to be used for long-term simulations.

The second stage (II) starts with the generation of representative sets of useful and believable control strategies of the complex system. The quality of the modelling done in the previous stage is subject to various extreme tests, as required in the industrial practice. Also the so-called long-term test is applied. The test discards strategies, without any further analysis, on the basis of the suspicion of being unfavourable in the long term, as shown by the simulation. Revisiting the model is possible at any time to achieve a sufficient validation with respect to the analysed problem.

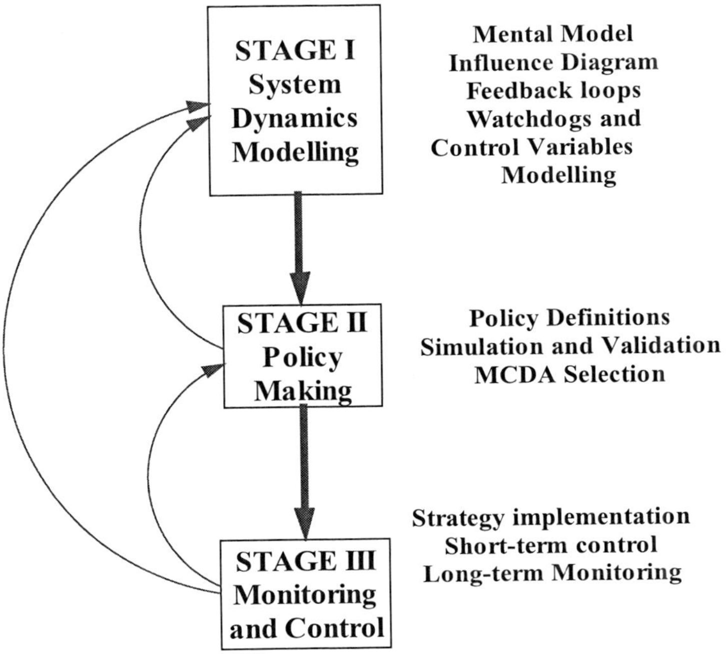

Figure 1. The three stages in ACM (Adative Control Methodology). MCDA=Multi-criteria Decision Aid.

A "good" strategy is then selected among the set of acceptable strategies passing all the tests. It is the place to incorporate group MCDA techniques into the decisional framework of Fig. 1 [4]. A 'good' compromise policy is selected among the set of acceptable strategies for the given list of criteria approved by the policy-makers. The selection in general takes into account the full time interval in which dynamic paths have been simulated with SD.

The third stage (III) in Fig. 1 is where real-time control sets in, after the selected policy has been actually implemented. Watchdogs are supposed to be monitored on a real-time basis, during the often very long periods representative of environmental issues. Regular updating of the policy is foreseen, e.g., every five years. Watchdogs may signal urgent short-term interventions. Use is made of the multiple branching opportunities in the ACM flowchart of figure 1 when revisiting policies, the model, or both, as appropriate.

In summary, ACM includes the following sequence of activities:

- Clear problem definition, and mental modelling are prerequisites in approaching the considered issue; a detailed influence diagram is developed on these premises; feedback mechanisms influencing or controlling the system are analysed;
- The control mechanisms, decision variables, the pursued objectives are decided upon;
- SD modelling and simulations are performed, e.g., with [3];
- Watchdogs are used as performance indicators and warning signals;
- The best performing control policies are selected ex-ante by means of MCDA techniques;

- Constant watchdog monitoring provides ex-post control of the chosen policies, and allows redoing on-line the previous steps in Fig. 1, when the model proves inadequate, or to change the control policy when performances do not fulfil expectations.

The fundamental concept of ACM control is now discussed more in detail. Three differences from the usual simulation approach with system-dynamics (SD) modelling may be pointed out:

- Firstly, in the usual SD control perspective the model and the actual system are kept apart during the planning phase in which policies are designed through simulations. In addition, the monitoring of further on-line behaviour of the real system does not provide any feedback from the controller to the system. This means that the planning is exclusively ex ante, without envisioning any policy change at later times. By contrast, in genuine control – and this is the same in engineered devices, although we point out here that human systems are not 'mechanical - the model ('controller') and the real system ('plant') are both part of a combined 'on-line' system.
- Secondly, SD control is not adaptive. Adaptive control implies that the controller can also change its structure on-line. Changes are made necessary by external events disturbing the system and the conditions of control. This implies that the influence diagram of the controlled system must be revised on-line. The authors have called 'control by structure' this advanced control procedure as transposed to human systems in ref. [5] to [7]. The primary objective is here to keep the system structurally stable, i.e. viable, which is actually the basic purpose of sustainable development. This control requires permanent information flows from observing the actual system evolution for being able to adapt the model and/or the policy-making. The structure of the system can then be modelled with increasing refinement. The design of adaptive policies is never definitely ascertained, however: unpredictable external events from outside the boundaries of the model may call for policy adaptations. Consequently, policy-making merely extrapolating past behaviours, or terminating right after implementation is often useless. Permanent follow-up is necessary to cope with future uncertainties.
- Thirdly, as said above, multiple goals in the search of control policies are often conflicting, and subject to compromises within the group of policy-makers, like searching compromises between economic necessities and sustainability needs. This supports the blending of systems techniques with group MCDA, applicable not only to off-line planning but also to on-line monitoring of the real system.

To perform control and monitoring in sustainable development thus requires being able to make long-term observations on the actual system behaviour, and to analyse how it reacts to the implemented policy, and to external 'disturbing' events.

2.2. Environmental Modelling with ACM

The author has produced several contributions to environmental modelling in which ACM has been applied in totality, or partially. An overview is provided in Table 1. These applications are shortly reviewed in this section. For obvious reasons ACM has never been used for real with the full control and monitoring scope of stage (III) (Fig. 1).

Ref. [7] provides a case where the ACM methodology has been applied by using fictive scenarios in which external events have required policy adaptations. This article investigates the application of a tax on fossil fuel, i.e., coal and natural gas, used for heating, lighting and transport in the residential sector. A SD model is developed to analyse the impact of the fuel tax on fuel consumption. The decision variables of the government are the CO_2 reduction objective, and the tax level. The tax is set at a sufficiently high level: the fuel price increase is expected to trigger investments in Rational-Use-of-Energy (RUE) technologies for reducing energy bills. Some care is taken to avoid excessive price increases, which may induce an interest rate increase reducing the borrowing capacity of RUE investors. Watchdogs are the actual CO_2 reductions, and the interest-rate evolution. External fictive events, like a sudden oil-price decrease, or an interest rate increase, are supposed to impact the system, and the actual evolution of fossil-fuel consumption. Policy-makers may have to adapt both the model, and the level of the tax according to the actual watchdogs evolution. Different alternatives for adapting the tax are investigated by means of group MCDA, assuming that several representative stakeholder groups (government, ecologist movements, etc.) are associated to the policy-making process.

Table 1. Overview of the author's ACM (Adaptive Control Methodology) applications in environmental modelling. The three ACM stages (I), (II), (II) are shown in Fig. 1.

Ref.	Topic	Control instrument	ACM stages
[7]	CO_2 emission reduction in the residential sector	Fuel tax	All three stages using future scenarios
[8]	Promotion of wind energy	Green Certificates (GC)	Planning stages (I) +(II)
[9]	CO_2-emission reduction in the electricity supply and demand sectors	Zero-Emission Certificates (ZEC)	Planning stages (I) +(II)
[10]	Fuzzy methodology for evaluating CO_2 reduction instruments	Emission permits Extendable to fuel taxes, GC, ZEC	Fuzzy-reasoning methodology is used for Planning (I)
[11]	CO_2 emission reduction in the residential sector	Fuel tax using the marginal abatement costs and the methodology from [10]	Planning stages (I) +(II)
[12]	Diffusion of decentralised Combined Heat Power (micro-CHP) in the residential sector	Economic incentives Resale of electricity to the grid	Planning stages (I) +(II)

Ref. [8] analyses the conditions in which wind energy can develop as a start-up technology in Belgium in 2002. Suitable promotion policies amplify the virtuous cycle of decreasing production costs with the installed capacity, starting from a very low capacity level. The SD simulations analyse the efficiency of a green-certificate (GC) market as a control instrument. Stakeholders are green electricity producers and regulators. These simulations only consider the ACM planning stages (I) and (II), ignoring the further downstream monitoring and control steps.

Decision variables are the dynamic quota of green electricity serving as a goal for the wind-energy growth, the GC price traded on a dedicated market, and the penalty value for electricity producers who do not produce a sufficient number of GC according to the imposed green-electricity production quota. The watchdogs are the gap between the green electricity goal and the actual production, the evolution of the cost of production of wind electricity, the amount of money perceived by the penalty payment, and the evolution of the electricity demand from the grid. An interesting observation appears in the model simulations: the development of windmills does always contribute to long-term CO_2-emission reductions. Although the long-term quota of green electricity is respected an important part of the production is used for increasing the total electricity demand. After a temporary flattening out of emissions has been observed, they continue to grow with demand at similar rates as before.

Ref. [9] investigates a new control instrument 'Zero Emission Certificates' (ZEC) for reducing CO_2 emissions. ZEC combine the advantages of green certificates (GC) and of CO_2 emission permits (EP), but they are designed such that well-known drawbacks are avoided, like compensating the increase in electricity demand without actually reducing carbon emissions for GC [8], or selling hot air with EP, rather than real carbon reductions.

The ZEC markets stimulates both supply and demand sides, and favours realistic and economically feasible reduction objectives. Only the ACM planning stages (I) +(II) are considered:

On the supply side electricity producers receive ZEC for any equivalent zero-emission MWh (=avoided emissions) obtained by substituting older fossil-fuelled power plant with more efficient modern combined-cycle gas-fuelled power plants, renewable energy sources, nuclear power plants, efficient combined heat-power systems, etc. The baseline for calculating the avoided emissions is chosen to be the old-fashioned coal-fuelled power plants emitting on average 800 kg CO_2 /MWh. In this way a universal benchmark is provided, and the past efforts of producers to reduce CO_2 emissions are rewarded. The ZEC may also be purchased on the specialised trading market, in case actual power-plant substitutions are more expensive than certificates acquisitions. ZEC trading thus stimulates the replacement of polluting power plants at the lowest possible costs, in a comparable way to EP trading.

In addition ZEC are used as a certification instrument for measuring the 'emission quality' of the electrical energy sold to distributors. The number of emitted ZEC give to the regulators the means of determining a quality objective in line with the average supply-side operating conditions. In this way a realistic and economically feasible CO_2 reduction objective can be worked out from the market data. This objective imposed to the distributors by the regulator is measured as a required percentage of emission-free electricity in the total energy distributed to final clients. This imposition is the ZEC equivalent to the GC quota. The ZEC quality indicator gives a further incentive to producers to orientate their power-plant substitution policy in line with the average emission conditions of the market.

On the demand side distributors are brought to respect the objective set by the regulator by encouraging electricity-demand reductions, and by preferring producers with a higher emission quality, i.e., less CO_2 emissions. Reduction-objective trespassing by distributors leads to penalty payments. The decision variables are the dynamic reduction objectives, and the penalty paid by the distributors. The watchdogs are the average zero-emission quality measuring the substitution of polluting plants by suppliers, and the demand evolution as an indicator of the efforts made for promoting RUE.

Ref. [10] presents a methodology paper to estimate the dynamic evolution of the marginal CO_2 reduction costs for different abatement techniques. The paper deals with emission permits (EP), but the fuzzy-reasoning technique may be used for all kinds of abatement instruments, including fuel taxes, GC, or ZEC (defined in Table 1). Fuzzy-rules techniques are used to reconcile the diverging opinions of experts, and to take into account the many uncertainties affecting the marginal abatement costs. Different experts' opinions are combined within the fuzzy rules by considering the experts' credibility levels.

Ref. [11] presents an improved and expanded carbon-tax SD model in the residential sector developing the model in ref. [7]. The new model takes into consideration the uncertainties on several variables, by applying the fuzzy-reasoning methodology developed in ref. [10]. Only ACM stages (I) and (II) are considered. The consumers' willingness to invest into RUE technologies under the influence of the carbon tax is modelled on the basis of experts' opinions regarding the marginal costs of abatement technologies, as in ref. [10]. The paper also shows how to reconcile with fuzzy reasoning the opinions of several experts regarding the evolution of loan interest rates, which could affect the willingness to invest into RUE. The decision variables are, as in ref. [7], the dynamic tax level, and the CO_2 reduction objective. Watchdogs are the investment rate into RUE technologies, the actual CO_2 reduction rate, and the evolution of the loan interest rates attached to RUE investment.

Ref. [12] examines the development of distributed technology in the residential sector, by means of small Combined Heat Power systems (micro-CHP) producing electricity and heat with high efficiency, and replacing traditional boilers in housings. Only stage I and II of the ACM have been analysed. The SD model shows that the efficient promotion of this technology requires adequate incentive policies from the authorities.

The decision variables of the model are the parameters characterising possible incentives: investment subsidies, energy price adaptations, reselling price of excess electricity, etc. The watchdog variables are the actual substitution rates of traditional boilers, and the natural-gas usage reduction, from which CO_2-emission reductions may be derived.

To complete this review of ACM applications, an additional case study is presented in the next section.

3. A Sustainability Case Study

3.1. Background

The purpose of this short presentation is to make clearer the concepts of control in ACM by means of a case study on various policies to reduce the fossil-fuel consumption in the residential sector. Public policies have the twofold objective of preserving non-renewable resources for future generations, and of capping the greenhouse gas emissions.

The residential sector in industrialised countries contributes to 40%-50% of all energy consumption - not including transport, not to be considered here. The role of public policies is to change the behaviour of residential consumers in their energy-usage patterns. Two kinds of economic instruments may be developed by policy-makers to reduce the CO_2-emissions in the residential sector:

- Additional charges, taxes;
- Incentives or direct subsidies.

3.2. The Initial Policies in 'NOW - 5 years'

In the energy-tax example handled in ref. [7], and shortly explained in sub-section 2.2, the authors had tried to avoid the difficulty of presenting real-time planning in ACM stage (III) by developing scenarios of plausible external events imposing a revision of the model, the policy options, or both. In the now presented case the author has chosen a different approach: it is assumed that initial decisions have been made five years back in time from the present time 'NOW' *without* using ACM, only on the basis of fully intuitive reasoning. It has been further assumed that the first in-depth monitoring and possible policy revision takes place 'NOW' *with* the use of quantitative modelling as recommended by ACM. This 'NOW' monitoring and policy-revision process is described in the following sub-section 3.3. The policy control instruments set in place at time 'NOW – 5 years' are the following ones:

I. Priority is given to the development on green electricity, mainly wind energy and biomass, by means of green certificates (GC) given to electricity production using renewable sources, and of direct incentives to private investors. The forecasted objectives is 15% at the 2020 time horizon;
II. Existing nuclear power plants representing about 50% of the local electricity production will be progressively shutdown after 40 years life-time, starting in 12 years and lasting 10 more years till the complete shut-down;
III. Newly constructed buildings must install solar panels on roofs or terraces to produce sanitary water; incentives are made available for adding such devices to more ancient buildings;
IV. Promotion campaigns regarding Rational Use of Energy (RUE) are started throughout the media (electricity and thermal demand reductions, e.g., use of low-consumption lighting devices; decrease of average room temperature; replacement of obsolete boilers; improved housing insulation, etc.). Subsidies and fiscal deductions are proposed to residential investors;
V. Budgets and fiscal incentives are proposed for advanced energy systems (micro-CHP systems, photovoltaic panels, etc.)
VI. A dedicated subsistence fund is established to subsidy poorer households for their heating oil or gas purchases.

Beyond the legal decision of abandoning nuclear production, three main decision variables are considered:

- The dynamically increasing minimum percentage of green electricity that must be provided year after year by the electricity suppliers, and the corresponding numbers of green certificates;
- The yearly budgets spent for promoting RUE in housings;
- The yearly budgets made available to investors for advanced energy systems.

The 'watchdog variables' to be monitored on a real-time basis are the following ones:

- The gap between the proposed percentage of green electricity, and the actual status of the green electricity production in TWh/year;
- The yearly average electricity and heat demands per households;
- The total yearly consumption of oil and gas, and the resulting CO_2 emissions;
- The number of successful subsidy files;
- The gap between the total domestic electricity demand and the domestic production, measuring electricity importations from abroad.

3.3. Monitoring 'NOW' the Initial Policies

The first monitoring phase ends 'NOW', i.e., five years after initiating the six policies described in section 3.2, and several conclusions are drawn:

I. Green energy, mainly electricity production by windmills and energy/heat production by biomass, is not developing as neatly as expected: the objective of 15% to be achieved in the long-term is not realistic in terms of surface occupation in a highly populated country. Furthermore, the low energy density and intermittency character of renewable sources bring difficulties in operating the electricity distribution network to cover base-load demand. In addition the development of windmills necessitates important back-up capacities, mainly gas-fuelled plants, to replace generators in periods without sufficient wind. The availability of on-shore windmills is only 30%;

II. The date of the first nuclear power plant (NPP) shutdown is in 7 years from NOW. This decision can only be maintained at the risk of a brutal increase of the electricity imports from abroad, i.e., an activation of the watchdog variable gap=(demand-domestic production), and also an important CO_2 increase rate, another watchdog variable;

III. Although the solar-panel law applicable to new buildings is generally applied, it has a rather small impact on the oil, and gas consumption. Relatively few owners of more ancient buildings introduce incentive files to the authorities, as this operation represents a lot of red tape, and the number of successful files is limited, another watchdog variable;

IV. Although there is much information and promotion around on climate change on the media, the impact on RUE is quite low: few households are indeed brought to change their consumption habits;

V. Although many files are introduced for getting financial support and/or fiscal incentives, the diffusion rates of advanced technologies remain limited to wealthier people because of the financial risk. Also, too optimistic cost decreases with scale/learning effect have been anticipated without the backing of quantitative models. Investment in new technology is sometimes used to create superfluous electricity or heat demands, and not so much to diminish the existing demand;

VI. Many people benefit from the support of the government to cap their oil and gas bills. Abuses exist: creation of superfluous demand; unwanted transfers from heat-producing boilers to electric heating radiators; etc.

As explained before in sub-section 3.1 the quantitative instruments of ACM (SD and MCDA) are assumed to assist the policy control process in stage (III) (Fig. 1).

3.4. Revisiting 'NOW' the Initial Policies

The quantitative simulation techniques of SD contribute in evidencing impossibilities resulting from too superficial analyses, or wishful thinking, like: "renewable energy can take over the lion share of fossil/nuclear sources".

Due to the diffuse and/or intermittency character of wind energy and biomass, and the near impossibility of developing other renewable energy sources, like hydro-electricity, the 15% objective of policy (I) will have to be revised downward in a considerable way. Examining all constraints, also the electricity network conditions, a maximum renewable energy production would be between 8% and 10% at the best.

Consequently the policy (II) of shutting down NPP starting in 7 years from 'NOW', has to be abandoned, or at least postponed, as its simulated application shows a considerable increase of electricity importations and/or of CO_2-emissions: fossil fuels will be mainly used for replacing the lost nuclear capacities (including possibly coal, while CO_2 sequestration methods are still far from operational).

Policies (III), (IV) are adequate, and in line with URE philosophy; efforts must be pursued. Required for a more successful promotion are: less 'red tape' for obtaining incentives, more carefully planned marketing campaigns, encouraging word-of-mouth from satisfied consumers, etc. Incentive procedures, and promoting budgets have to be revisited, using also the accumulated experience from the last five years.

SD models are prepared to analyse the diffusion capabilities of new technologies, and of URE behaviours. These simulations show the crucial importance of the word to mouth and advertisement on the households. Fiscal advantages or energy-price reductions also appear as important economic drivers, as illustrated in [12] for the diffusion of micro-CHP.

The most policy revisions affect policies (V) and (VI): the priorities for attributing subsidies have to be completely revisited; energy taxes are also considered now for complementing subsidies. The analyses are made with SD modelling, and MCDA is used to rank proposed policies according to decreasing priorities. The analysis takes into account the constraint of limited budgets, and considers that part of the fiscal incomes has been attributed by priority to the URE policies described under the headings (III) and (IV).

Regarding the use of subsidies encouraging advanced technologies, it appears that the attempt to introduce them into the broad public would be premature; also the learning effects are affected by very large uncertainties, which counteract the large-scale deployment of these technologies. The best example is here photovoltaic electricity, the production cost of which is today several times above the average production cost of electricity from the centralised network.

The use of the subsidy model of ref. [12] for simulating the penetration of these new technologies in households indeed confirms this observation. To be efficient – in the most favourable cases - the SD model shows that subsidies must represent an important percentage of the initial investment during many years, i.e., a large part of the total available budget. In addition high diffusion rates suppose sufficient investment-cost decreases with the installed capacity of the substitute technology, and favourable energy-price evolutions. Enormous uncertainties are attached to these external influence factors on which policy-makers have no control. The analysis indicates that the highest priority for allocating budgets must be given to R&D for studying efficiency improvements, economic-scale effects, performing storage systems, etc. By comparison a general tax model applied to all fuel consumption would most influence the general-public behaviour [11].

In summary, the MCDA analysis of different options would provide the following quite sensible recommendations for the next 5 years, till the next periodical revision:

- Public money should be reserved in priority to URE incentives and promotion campaigns;
- Advanced technologies – photovoltaic panels, micro-CHP, etc. - should be first brought to more maturity and cost reductions by increasing R&D budgets, and limiting the first introductions of these technologies to prototype projects;
- A general energy tax system on fossil fuels should be given the preference over incentive schemes, fitted with some cost-reduction provisions for those consumers ready to implement URE measures in their homes. In this way poorer households could improve their conditions of living and reduce at the same their energy bills, e.g., by improving the home insulation, which would make unnecessary the use of the special fund in policy (VI).

4. Conclusion

This article stresses the importance of using a suitable quantitative framework for policy designing in the complex issues of sustainable development, an important ethical challenge in the 21st century.

This type of prospective framework provided by the Adaptive Control Methodology (ACM) confirms the observation that adaptive, and flexible responses to changes in the environment are indispensable complements to planning. The presence of important uncertainties and simplifications in the models, the possibility of unpredictable external shocks cannot limit the policy-making process to off-line planning. All along the chosen time-horizon, policy-makers have to proceed to one-line adaptations and revisions of the chosen policy measures and their modelling. Of course the process of monitoring and policy revision (ACM stage III in Fig. 1) must be permanently fed through an adequate information flow. It is a condition for correctly monitoring watchdog-variables relevant for the case. This kind of tool is adequately called a Policy Support System (PSS), rather than a Decision Support System (DSS). A basic difference between both support systems is related to the aspects of the policy follow-up. A DSS designs a priori solutions, giving no or only trivial care to the subsequent execution. By contrast, a PSS combines permanent monitoring, and reengineering tasks, with the initial policy design. Another important part of a PSS is to foresee negotiations between multiple agents, and it why group MCDA is an inseparable part of ACM in order to implement and adapt policies in the course of time. Planning under uncertainty, monitoring and controlling complex socio-economic systems are never-ending tasks.

References

[1] Kunsch, P.L., Theys, M., Brans, J.P. The importance of systems thinking in ethical and sustainable decision-making, *Central European Journal of Operations Research*, 2007 15/3, 253-269.

[2] Sterman, J.D. *Business Dynamics*. Systems Thinking and Modeling for a Complex World, Boston, MA: Irwin McGraw-Hill, 2000.

[3] Vensim ® Ventana® Simulation Environment Version 5, User's Guide, Harvard, MA: Ventana Systems, Inc., 2002.

[4] Figueira, J., Greco, S., Ehrgott, M. (eds.) *Multiple Criteria Decision Analysis*, State of the Art Surveys, New York: Springer Science+Business Media, Inc., 2005.

[5] Brans J.P., Macharis, C., Kunsch P.L., Chevalier A. and M. Schwaninger Combining multicriteria decision aid and system dynamics for the control of socio-economic processes. An iterative real-time procedure, *European Journal of Operational Research*, 1998 109(2), 428-441.

[6] Brans, J.P., Kunsch, P.L., and Mareschal, B. Management of the future. *A system dynamics and MCDA approach,* in, D. Bouyssou, E. Jacquet-Lagrèze, P. Perny, R. Slowinski, D. Vanderpooten, Ph. Vincke (Eds.), Special volume dedicated to Professor Bernard Roy, Dordrecht, The Netherlands: Kluwer Academic Publishers, 483-502, 2002.

[7] Kunsch, P.L., Springael, J., and Brans, J.P. An adaptive multicriteria control methodology in sustainable development - Case study: a CO_2 ecotax, in, A. Colorni, M. Parrucini, and B. Roy (Eds.), Advances in Multicriteria Decision Aid, Report of the Commission of the European Communities, Joint Research Center, Ispra, Italy: EUR-report 19808, 145-161, 2001.

[8] Kunsch, P.L., Álvarez-Nóvoa, R., and Springael, J. Simulation d'un marché de certificats verts pour la promotion de l'énergie éolienne en Belgique (Simulation of a green-certificate market for promoting wind energy in Belgium), INFOR, Canadian *Journal of Operational Research and Information Processing*, 2002 40/3, 241-258.

[9] Kunsch, P.L., Springael, J., and Brans, J.P. The Zero-Emission Certificates: A novel CO_2-pollution reduction instrument applied to the electricity market, *European Journal of Operational Research*, 2004 153/2, 386-399.

[10] Kunsch, P.L., and Springael, J. A fuzzy Methodology for evaluating a market of tradable CO_2-permits, in, *Energy and Environment*, R. Loulou, J.-Ph. Waaub, G. Zaccour (Eds.), Gerad 25th Anniversary Series, New York: Springer Science+Business Media, Inc., Chapter 6, 149-173, 2005.

[11] Kunsch, P.L., and Springael, J. Simulation with System Dynamics and fuzzy control of the reduction of CO_2-emissions in the residential sector, *European Journal of Operational Research*, 2008 185, 1285-1299.

[12] Ben Maalla, E.M., and Kunsch, P.L. *Simulation of micro-CHP diffusion by means of system dynamics, Energy Policy*, Special Issue on Operational Models and Methods in the Energy Sector, 2008, Article in Press, available on line 5 March 2008.

In: Environmental Modelling: New Research
Editor: Paul N. Findley, pp. 151-185

ISBN: 978-1-60692-034-3

Chapter 7

PREDICTION OF SEDIMENT SOURCE AREAS WITHIN WATERSHEDS AS AFFECTED BY SOIL DATA RESOLUTION

Xixi Wang[1,*], Javier Garza[2], Miles Whitney[3], Assessfa M. Melesse[4] and Wanhong Yang[5]

[1]Assistant Professor, P.E., Hydrology and Watershed Management Program, Engineering & Physics Department, Tarleton State University, Box T–0390, Stephenville, TX 76402

[2]Associate Professor and Head, Department of Mathematics, Tarleton State University, Box T–0470, Stephenville, TX 76402

[3]Undergraduate Research Assistant, Hydrology and Watershed Management Program, Engineering & Physics Department, Tarleton State University, Box T–0390, Stephenville, TX 76402

[4]Associate Professor, P.E., D.WRE, Department of Environmental Studies, Florida International University, Miami, FL 33199

[5]Associate Professor, Department of Geography, University of Guelph, Guelph, Ontario, Canada N1G 2W1

Abstract

Distributed watershed models, such as the Soil and Water Assessment Tool (SWAT), have proven to be an important means for tackling problems related to nonpoint source pollution. These models require soil data as one of the minimal inputs because soil controls the runoff mechanism of a watershed. The U.S. Department of Agriculture Natural Resources Conservation Service generated two national soil databases, namely the State Soil Geographic (STATSGO) and Soil Survey Geographic (SSURGO) databases, which were designed to be used for river basin resource monitoring, planning, and management. STATSGO has a county level spatial resolution, whereas, SSURGO has a farm level resolution. These two databases have been widely used as the best alternatives for watersheds where site specific soil data are

* E-mail address: xxqqwang@gmail.com. Tel. (254) 968–9164. Corresponding Author: Xixi Wang, Hydrology and Watershed Management Program, Engineering & Physics Department, Tarleton State University, Box T–0390, Stephenville, TX 76402.

not available, which is a common case. However, in the literature, information is scarce regarding affects of using one of these two databases over another on modeling sediment processes. Thus, the objective of this study was to evaluate affects of using STATSGO versus SSURGO as an input on SWAT simulated sediment yields at the outlet of, and sediment source areas within, a watershed. The evaluation was conducted in one North Dakota watershed and one Texas watershed. Obviously, these two watersheds have distinctly different climatic and hydrologic conditions. The results indicated that for each of the study watersheds, the predicted total sediment yields at the outlet using STATSGO might be comparable with the corresponding predicted values using SSURGO, the predicted source areas, however, could be very different. In addition, the prediction discrepancies as a result of using one dataset over another were larger in the Texas watershed. This might indicate that the simulation of a watershed with streamflows mainly generated from rainfall runoff is likely to be more sensitive to soil data resolution than the simulation of a watershed with streamflows predominantly generated by snowmelt runoff.

Keywords: NPS, data resolution, sediment simulation, soil data, source areas, SWAT

Introduction

Soil is important in controlling the runoff mechanism of watersheds. This is particularly true for watersheds where the saturation excess runoff prevails (Dunne et al., 1975). Soil within a watershed is seldom uniform and usually exhibits some degrees of spatial variability. For soil survey purposes, the variability is signified using a number of soil map units (SMUs), each of which is assumed to consist of soils with similar properties (e.g., hydraulic conductivity). A SMU has one dominant soil and may include one component soil or several. Thus, details of the soil variability that can be presented by a soil map are dependent on the total number of SMUs (i.e., the spatial scale of the map) (Western et al., 2002).

Using the soil maps produced from different intensities and scales of mapping, the U.S. Department of Agriculture (USDA) Natural Resources Conservation Service (NRCS; formerly the Soil Conservation Service or SCS) generated two datasets, namely State Soil Geographic (STATSGO) and Soil Survey Geographic (SSURGO) databases, which can be used for river basin resource planning, management, and monitoring (USDA–NRCS, 2005). Because these two databases are available for the entire land mass of the United States and provide information that is sufficient for general watershed scale studies, they have been commonly used as the best alternatives for watersheds where site specific data are not available. Released earlier and with a smaller file size, STATSGO was widely used to set up watershed models (e.g., Chu et al., 2004; Wang and Melesse, 2005). However, given that SSURGO has a much higher spatial resolution and presents more detailed information, researchers (e.g., Wang and Melesse, 2006; Muleta et al., 2007) have an increasing interest in using SSURGO as an input of watershed models. For this reason, the information is very necessary on affects of using one of these two databases over another on modeling watershed processes (e.g., sediment transport). Nevertheless, such information is very scarce in the literature.

To facilitate the use of SSURGO within the Soil and Water Assessment Tool (SWAT; Arnold et al., 1998), Peschel et al. (2006) developed an ArcView extension called "SSURGO SWAT 2.0." This extension can be used to generate a user soil database in the format required by ArcView Interface for SWAT (Di Luzio et al., 2002) by querying three attribute tables of

SSURGO (ver. 1.0), namely map unit, component, and horizon (USDA–NRCS, 2005). In addition, the authors conducted a case study for the Upper Sabinal River watershed located in Central Texas. The results indicated that using the SSURGO data tended to give higher water yields than using the STATSGO data, which the authors attributed to the significantly larger values of saturated hydraulic conductivity presented by the SSURGO data. This explanation seems vague because a larger saturated hydraulic conductivity means that a greater proportion of the rainfall might be infiltrated into ground, leading to a smaller direct runoff. To make this explanation convincible, the predicted total streamflows may need to be separated into the corresponding direct and base flows. Provided that this explanation is reasonable, the ratios of the predicted base flows to the corresponding total streamflows using the SSURGO data are expected to be larger than those using the STATSGO data. As a result, the predicted sediment yields using the SSURGO data are expected to be smaller than the corresponding predicted values using the STATSGO data, which was not verified by the authors.

In a separate study, Wang and Melesse (2006) evaluated the affects of using the STATSGO versus SSURGO data on the predicted total streamflows in the Elm River watershed located in Eastern North Dakota. Because the SSURGO SWAT 2.0 extension is incapable of processing the upgraded SSURGO (ver. 2.0) data, a computer program was written in Microsoft Visual Basic® and used to generate the user soil database in the format required by SWAT. The authors found that while the predicted total streamflows at the watershed outlet were comparable, the low flows were more accurately predicted by using the STATSGO than SSURGO data. In addition, the discrepancies between the predicted discharges using the STATSGO data and the corresponding predicted values using the SSURGO data tended to be larger at upstream locations than at those further downstream within the study watershed. Further, the affects of using one of these two databases over another might be more noticeable in the areas where overland runoff processes are predominant than in the areas where channel routing processes become more important. However, the authors did not examine whether the similar affects would exist for sediment simulation, i.e., the discrepancies of predicted sediment yields from using one of these two databases over another are expected to be larger for contribution areas (i.e., subbasins in SWAT) dominated by overland runoff processes than for those dominated by channel routing processes.

The objective of this study was to evaluate influences of using STATSGO versus SSURGO on sediment prediction. Given the aforementioned findings of the previous studies (Peschel et al., 2006; Wang and Melesse, 2006), this study hypothesized that using one of these two soil databases over another would give distinctly different spatial distributions of sediment contribution areas within a watershed. Herein, the spatial distribution is defined by overlaying a map showing the predicted sediment yields with another map showing the predicted sediment contribution probabilities. For a sediment contribution area, the sediment contribution probability is computed as the percentage of the evaluation days with a nonzero sediment yield. In addition, this study hypothesized that the influences would be dependent on climatic and hydrologic conditions of the watershed because of the direct effects of these factors on soil erosion and sediment transport (Juracek and Ziegler, 2007). This study was conducted in the upper portions of the Elm River watershed located in eastern North Dakota (Figure 1) and of the Cowhouse Creek watershed located in north central Texas (Figure 3).

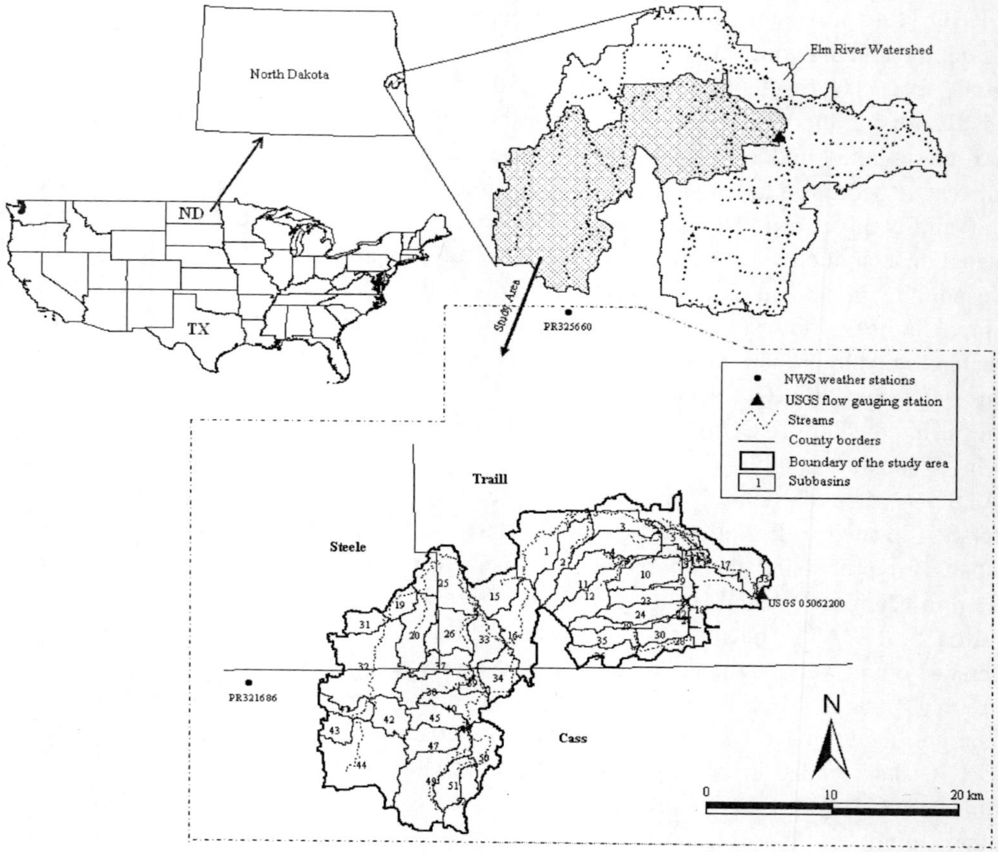

Figure 1. Map showing the location of the Elm River watershed, the boundaries of the NDERW study area and its 53 subbasins, and the two National Weather Service (NWS) stations and the U.S. Geological Survey (USGS) flow gauging station where the data were used in this study.

Description of SWAT

SWAT is a physically based, watershed-scale, continuous-time, distributed-parameter hydrologic model that uses spatially distributed data on topography, land use, soil, and weather for hydrologic modeling and operates on a daily time step (Arnold et al., 1998). Based on topography, SWAT subdivides a watershed into a number of subbasins for modeling purposes. Each subbasin delineated within the model is simulated as a homogeneous area in terms of climatic conditions, but additional subdivisions are used within each subbasin to represent different soils and land use types. Each of these individual areas is referred to as a hydrologic response unit (HRU) and is assumed to be spatially uniform in terms of soils, land use, and topography. SWAT is composed of three major components, namely subbasin, reservoir routing, and channel routing, each of which includes several subcomponents. For example, the subbasin component consists of eight subcomponents: hydrology, weather, sedimentation, soil moisture, crop growth, nutrients, agricultural management, and pesticides. The hydrology subcomponent includes surface runoff, lateral subsurface flow, percolation, ground water flow, snowmelt, evapotranspiration, transmission losses, and ponds. Detailed descriptions of the methods used in modeling these components and subcomponents can be found in Arnold et al. (1998), Srinivasan et al. (1998), and Neitsch et al. (2002b). In this study, the SCS runoff curve number method, with the SCS curve number adjusted according to soil moisture conditions (Arnold et al., 1993), was used to estimate surface runoff; the Hargreaves method (Hargreaves and Samani, 1985) was used to estimate potential evapotranspiration (PET); and the Muskingum method (Chow et al., 1988) was used for channel routing.

SWAT uses four storage volumes to represent the water balance in each HRU in the watershed: snowpack, root zone (0 to 2 m), shallow aquifer (2 to 20 m), and deep aquifer (> 20 m). The root zone can be subdivided into multiple layers. The excess water available after accounting for initial abstractions and surface runoff, using the SCS curve number method, infiltrates into the soil. A storage routing technique is used to simulate the flow through each soil layer. SWAT directly simulates only the saturated flow and assumes that water is uniformly distributed within a given layer. The unsaturated flow between layers is indirectly modeled using depth distribution functions for plant water uptake and soil water evaporation. Downward flow occurs when the soil water in the layer exceeds field capacity and the layer below is not saturated. The rate of downward flow is governed by the saturated hydraulic conductivity. Lateral flow in the root zone is simulated using a kinematic storage routing technique that is based on slope, slope length, and saturated conductivity. Upward flow from a lower layer to the layer above is regulated by the soil water to field capacity ratios of the two layers. Percolation from the bottom of the root zone is recharged into the shallow aquifer.

Water movement in the soil is impacted by the soil temperature. In SWAT, the daily average soil temperature is calculated at the soil surface and the center of each soil layer. The temperature of the soil surface, which may be related to wind speed and direction because wind affects the sensible heat exchanged between the air and the soil surface (Condie and Webster, 1997), is modeled by SWAT as a function of snow, plant and residue covers, bare soil surface temperature, and the previous day's soil surface temperature. The temperature of a soil layer is modeled as a function of the surface temperature, mean annual air temperature, and the depth in the soil at which a variation in temperature due to changes in climatic conditions no longer occurs. This depth, referred to as the damping depth, is dependent upon the bulk density and the soil water content.

SWAT classifies precipitation as either rain or snow based on the mean daily air temperature and a specified boundary temperature (i.e., snowfall temperature); the precipitation is classified as snow when the mean daily air temperature is less than the boundary temperature and as rain when the air temperature is greater. The water equivalent of the snow precipitation is then added to the snowpack. The snowpack will increase with additional snowfall and decrease with snowmelt and sublimation. Snowmelt is controlled by the air and snowpack temperatures, the melting rate, and the areal coverage of snow. The snowpack temperature on a given day is estimated as the weighted average of that day's mean air temperature and the snowpack temperature on the previous day. The weighting includes a specified lag factor, which accounts for the snowpack density, snowpack depth, exposure, and other factors affecting the snowpack temperature. The snow melting rate is allowed to have a seasonal variation, with the specified maximum and minimum values occurring on the summer and winter solstices, respectively.

The areal coverage of snow has a good correlation with the amount of snow present in a watershed of interest at a given time because other factors that contribute to variations in the snow coverage, such as drifting, shading, and topography, are usually similar from year to year (Anderson, 1976). This correlation is expressed in SWAT as an areal depletion curve, which is used to describe the seasonal growth and recession of the snowpack as a function of the amount of snow present in the watershed. The areal depletion curve requires a threshold depth of snow above which there will always be 100 percent cover. The threshold depth depends on factors such as vegetation distribution, wind loading and scouring of snow, interception, and aspect and is unique to the watershed. This snow accumulation and melt phenomenon is modeled using seven parameters in SWAT that are discussed in detail by Neitsch et al. (2002a, b).

For watersheds with a high topographic relief (e.g., Fontaine et al., 2002), the snowpack distribution varies with topographic elevation. In this regard, SWAT uses elevation bands to distribute temperature and precipitation when simulating snowmelt. However, because the Elm River watershed has a very low topographic relief (Houston Engineering, 2001), the precipitation and temperature within the watershed vary negligibly with topographic elevation. Hence, the effects of elevation on the snowpack distribution and the snowpack temperature were not considered in this study. In addition, a study conducted by Wang and Melesse (2005) in the Wild Rice River watershed in northwestern Minnesota indicated that three of the seven parameters, the maximum snowmelt factor (variable SMFMX), snowpack temperature lag factor (variable TIMP), and base temperature above which snowmelt is allowed (variable SMTMP), were sensitive to the snowmelt. Because the Elm River watershed is adjacent to the Wild Rice River watershed, it was reasonably assumed that these three parameters would also be sensitive to the snowmelt in this study area and thus were taken as the calibration parameters in addition to others. In contrast, these parameters were not adjusted for the study area within the Cowhouse Creek watershed, where there was no snowpack accumulated despite the sporadic snow in winter.

SWAT simulates sediment yield using the Modified Universal Soil Loss Equation (MUSLE; Williams, 1995). MUSLE predicts the amount of eroded soils in a HRU that would be delivered into the channel of the inclusive subbasin. In the channel network, sediment transport is modeled as a function of two processes, deposition and degradation, operating simultaneously in the reach (Bagnold, 1977; Williams, 1980). The maximum amount of sediment that can be transported from a reach segment is a function of the peak channel velocity. SWAT provides two options to compute deposition and degradation (Neitsch et al., 2002b). The first option is to use the same channel dimensions for the entire simulation, whereas, the second option is to simulate downcutting and widening of the channel and update channel dimensions throughout the simulation. In this study, the first option was used for both study areas.

STATSGO versus SSURGO

The SSURGO soil database was designed by the USDA--NRCS (1995) to provide farm-level spatial resolution for farm and ranch, landowner/user, township, county, or parish natural resources planning and management. Data for SSURGO are collected and archived in the

U.S. Geological Survey (USGS) 7.5-minute topographic quadrangle units and distributed as a complete coverage for a soil survey area that usually consists of 10 or more quadrangle units. The adjoining 7.5-minute units are matched within the survey areas. Peschel (2006) found that SSURGO does not assign a classification for soils perennially covered by water and suggested it would be reasonable to assume that these areas have the same soil classifications as adjacent areas. For the study area within the Elm River watershed, the soils for 9 areas with a size of less than 1 ha were perennially covered by water, whereas, for the study area within the Cowhouse Creek watershed, the soils for 3024 areas with a size of up to 5 ha were perennially covered by water. The soils in these areas were assumed to have the same classifications as the soils in the nearest areas. This assumption is likely to have a negligible influence on the study results because of the small sizes of, and the indirect contact between runoff and soils within, these areas. In contrast, the STATSGO database, designed by the USDA-SCS (1993), has a county-level resolution. Data for STATSGO are collected in the USGS 1:250,000-scale, 1- by 2-degree topographic quadrangle units and merged and distributed as statewide coverages. Features are edge-matched between states. The STATSGO database is generally used for regional, multistate, river-basin, state, and multicounty resource planning, management, and monitoring.

Both STATSGO and SSURGO could be used as the input of SWAT for the purpose of hydrologic modeling. However, depending on the modeling objective and accuracy, each has advantages as well as disadvantages. Because of its low resolution, STATSGO may assign a single classification to areas that actually have several different soil types. As a result, the individual effects of these misclassified soils on the model simulation may not be taken into account by a SWAT-STATSGO model. In contrast, SSURGO has a higher spatial resolution and will include the soils that are misclassified in STATSGO. As a result, the individual effects of these soils would be considered by a SWAT-SSURGO model. Nevertheless, the trade-off involved in using SSURGO rather than STATSGO as the model input is that more HRUs and their associated parameters must be determined, mandating a more extensive model calibration. Further, the larger number of parameters in a SWAT-SSURGO model may increase the model prediction uncertainty (Gupta et al., 1999). Given that soil mainly affects overland hydrologic processes (Wang and Melesse, 2006) and that the relative importance of channel processes to overland processes increases when moving downstream (Hill and Neary, 2005), it may be that for a watershed, the differences between the streamflows predicted by a SWAT-STATSGO model and the corresponding values predicted by the complementary SWAT-SSURGO model would be smaller downstream than at an upstream point within the same watershed (Wang and Melesse, 2006). Thus, when the streamflows at downstream points, where the streamflows could be predicted either by a SWAT-STATSGO model or the complementary SWAT-SSURGO model with a similar accuracy, are of interest, STATSGO may be superior to SSURGO due to its smaller file sizes and the ease with which the data can be processed by AvSWAT.

Study Area within the Elm River Watershed

The 51,538 ha drainage area upstream of station USGS 05062200 within the 114,700 ha Elm River watershed in eastern North Dakota (Figure 1), designated NDERW for description purposes, was taken as one of the two study areas. NDERW covers portions of three North

Dakota counties: Cass, Steele, and Traill. Based on the LULC data from the U.S. Environmental Protection Agency (USEPA), the land use within the study area consists of almost entirely agriculture (98.6%), and the rest is composed of range grasses (0.51%), transportation (0.22%), nonforested wetlands (0.20%), medium-density residential areas (0.19%), deciduous forest (0.12%), waters (0.10%), and commercial areas (0.06%). While the LULC data were developed from 1970s and 1980s aerial photographic surveys combined with land use maps and surveys (USEPA, 2003), they are effectively up to date because changes have been negligible in the land use types for NDERW in the past few decades (Stoner et al., 1993). The soils in NDERW are dominated by clay and clay loams, with an infiltration rate ranging from 0.5 to 250 mm/hr. The soils are classified into 16 types in the STATSGO data but are categorized into 88 types in the SSURGO data. For illustrative inspection purposes only, the STATSGO and SSURGO delineations for the soils within NDERW are shown in Figures 2(a) and 2(b), respectively. Although it is not feasible to discern the soil types from these figures due to their scale, it is evident that SSURGO is far more detailed than STATSGO. NDERW has a very low topographic relief and is characterized by broad, flat alluvial floodplains, river terraces, and gently sloping uplands (Moore Engineering Inc., 1982); its local relief is less than 5 m and global relief is only up to 150 m, with the elevation ranging from 256 to 390 m. The precipitation and temperature within NDERW vary negligibly with topographic elevation.

The National Weather Service (NWS) National Climate Data Center (NCDC) collects data on daily precipitation at stations PR321686 and PR325660, which abut NDERW (Figure 1). In addition, NCDC collects data on daily minimum and maximum temperatures at station PR325660. The numbers in these labels signify the 6-digit NWS COOP IDs (cooperative station identifiers) for these stations. An examination of the data revealed that the records for the period from 1 October 1981 to 30 November 1986 were complete except for April 1983 and February 1984, when the daily precipitation data were missed for station PR321686. The missing values were arbitrarily filled with the corresponding values measured at station PR325660; this substitution is expected to exert only a limited influence on the simulation results in this study because of the low percentage (less than 3.5%) of missing data. The data indicated an annual average precipitation of 470 mm, 25 percent of which (118 mm) was in the form of snowfall. The annual average daily temperature ranged from -40°C in winter to 46°C in summer, with a mean of 4.8°C. During the three summer months (June through August), the temperature could be 32 °C or higher for 17 days; during the three winter months (December through February), 54 days could have a temperature below -18°C. Overall, the study area had a cool, windy, and dry to moderately wet climate (Moore Engineering Inc., 1982). Such climate conditions may make the Hargreaves method an appropriate approach for PET estimation (Martinez-Cob and Tejero-Juste, 2004). Further, the Hargreaves method was used in this study because the additional data (e.g., net solar radiation) required by the other two SWAT provided methods, Penman-Monteith (Penman, 1956; Monteith, 1965) and Priestley-Taylor (Priestley and Taylor, 1972), were unavailable.

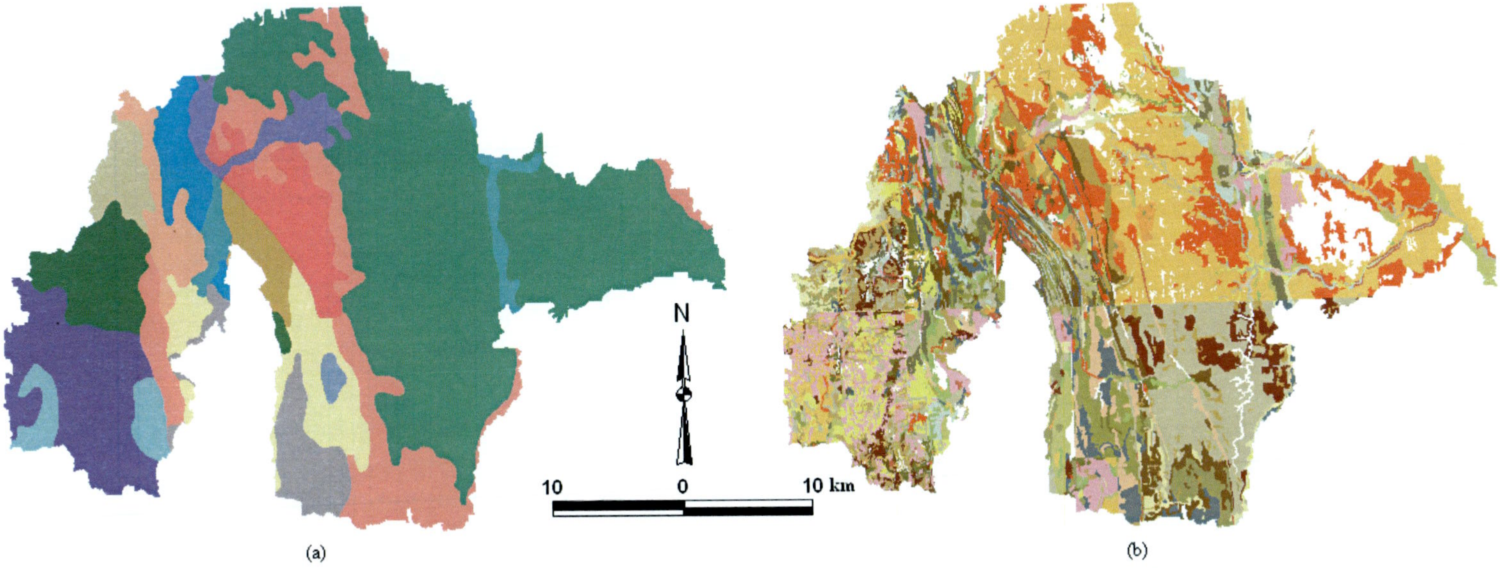

Figure 2. Map showing classifications of the soils in the Elm River watershed, located in eastern North Dakota, in the two U.S. Department of Agriculture (USDA) Natural Resources Conservation Service (NRCS) soil databases of (a) the State Soil Geographic (STATSGO) with a county-level spatial resolution, and (b) the Soil Survey Geographic (SSURGO) with a farm-level spatial resolution. The purpose is to illustrate that the SSURGO data are much more detailed than the STATSGO data rather than to show the discernable classifications.

The USGS monitored daily streamflows at station USGS 05062200 from 1 October 1955 to 31 August 1986. The Elm River was a seasonal stream, with near-zero flows in fall and winter, intermittent flows in summer, and continuous flows in spring (Moore Engineering Inc., 1982). The streamflows in summer were predominantly generated from rainfall runoff, but the streamflows in spring were predominantly generated from melting snow. Although the USGS recorded the daily streamflows for the period from 1 October 1955 to 30 September 1963, more than 90 percent of the days had zero or near-zero flows. Further, the daily flows in spring for these years were less than 2.0 m^3/s, except that the daily peaks of 4.4 m^3/s and 9.3 m^3/s occurred in 1960 and 1962, respectively. The values from 1 October 1963 to 30 September 1980 were missed, and the daily flows from 1 October 1980 to 30 September 1981 were zero or near zero. In contrast, the water years from 1982 to 1986 had relatively larger daily flows in spring and summer. While the values in fall and winter for these five water years were unavailable, the daily flows for these two seasons were usually very low and could be assumed to be zero (Wiche and Williams-Sether, 1997; G. J. Wiche, 2005, personal communication).

Herein, a water year was defined as from December to November to more accurately reflect the hydrologic characteristics of the NDERW study area (Moore Engineering Inc., 1982) and to include all four seasons for convenience of interpreting the modeling results. The available data indicated that at station USGS 05062200, the daily streamflows before 1982 ranged from 0 to 9.3 m^3/s, with a mean of 0.06 m^3/s, and the daily streamflows from 1982 to 1986 varied from 0 to 20.7 m^3/s, with a greater mean of 1.12 m^3/s. The data also indicated that the daily peaks, with a magnitude ranging from 6.4 to 20.7 m^3/s, always occurred in spring. Hence, the data on daily streamflows for the water years from 1982 through 1986 were used to calibrate and validate the SWAT models, which had a modeling domain covering the upper 45 percent of the Elm River watershed.

Study Area within the Cowhouse Creek Watershed

The 117,845 ha drainage area upstream of station USGS 08101000 within the 188,715 ha Cowhouse Creek watershed in north central Texas (Figure 3), designated TXCCW for description purposes, was taken as the second study area. TXCCW covers portions of five Texas counties: Comanche, Coryell, Hamilton, Lampasas, and Mills. Based on the National Land Cover Data 2001 (NLCD 2001; Homer et al., 2004) from USEPA, the land use within the study area consists of almost entirely range brush (55.0%) and range grasses (40.0%), and the rest is composed of forest (4.4%), waters (0.4%), and agriculture (0.1%). NLCD 2001 distinguishes 16 different classes (USDA-NRCS, 1995) and is the best available land use data covering the entire study area. The soils in TXCCW are dominated by clay and sandy clay, with an infiltration rate ranging from 0.8 to 332 mm/hr (Mace et al., 1999; TCEQ, 2007). The soils in TXCCW are classified into 7 types in the STATSGO data but are categorized into 91 types in the SSURGO data. For illustrative inspection purposes only, the STATSGO and SSURGO delineations for the soils are shown in Figures 4(a) and 4(b), respectively. As for NDERW, although it is infeasible to discern the soil types from these figures due to their scale, it is evident that SSURGO is far more detailed than STATSGO. TXCCW also has a very low topographic relief; its local relief is less than 5 m and global relief is only up to 340 m, with the elevation ranging from 150 to 490 m. The Cowhouse Creek has an average sinuosity of 1.3 and is characterized by broad flat alluvial floodplains, river terraces, and lower Cretaceous Walnut Clay with adjacent ridges of the Edwards-Duck Creek-Comanche Peak limestone (Nordt, 2004).

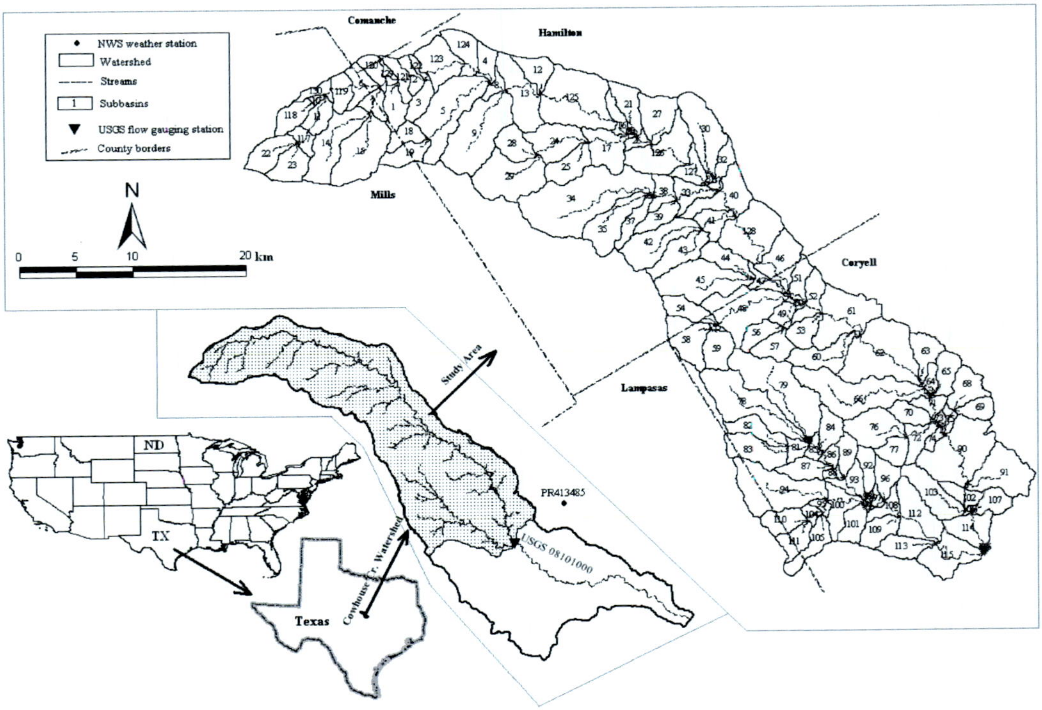

Figure 3. Map showing the location of the Cowhouse Creek watershed, the boundaries of the TXCCW study area and its 130 subbasins, and the National Weather Service (NWS) station and the U.S. Geological Survey (USGS) flow gauging station, where the data were used in this study.

NCDC collects data on daily precipitation and minimum and maximum temperatures at station PR413485, which abuts TXCCW (Figure 3). An examination of the data revealed that the records for the period from 1 January 1980 to 31 December 2006 had only 32 missing values for precipitation and had no missing value for temperatures. The missing values were generated by the weather generator that is a component of the SWAT software package (Neitsch et al., 2002b). This is expected to exert only a limited influence on the simulation results because of the low percentage (less than 0.32%) of missing data. The data indicated an annual average precipitation of 865 mm. The annual average daily temperature ranged from -20°C in winter to 45°C in summer, with a mean of 19°C. During the three summer months, the temperature could be 32 °C or higher for 75 days; during the three winter months, 30 days could have a temperature below zero. Overall, TXCCW had a cool, windy, and wet climate (Greer, 2005). Again, such climate conditions may make the Hargreaves method an appropriate approach for PET estimation (Martinez-Cob and Tejero-Juste, 2004). In addition, the Hargreaves method was used for PET estimation because the additional data (e.g., net solar radiation) required by the other two SWAT provided methods, Penman-Monteith (Penman, 1956; Monteith, 1965) and Priestley-Taylor (Priestley and Taylor, 1972), were

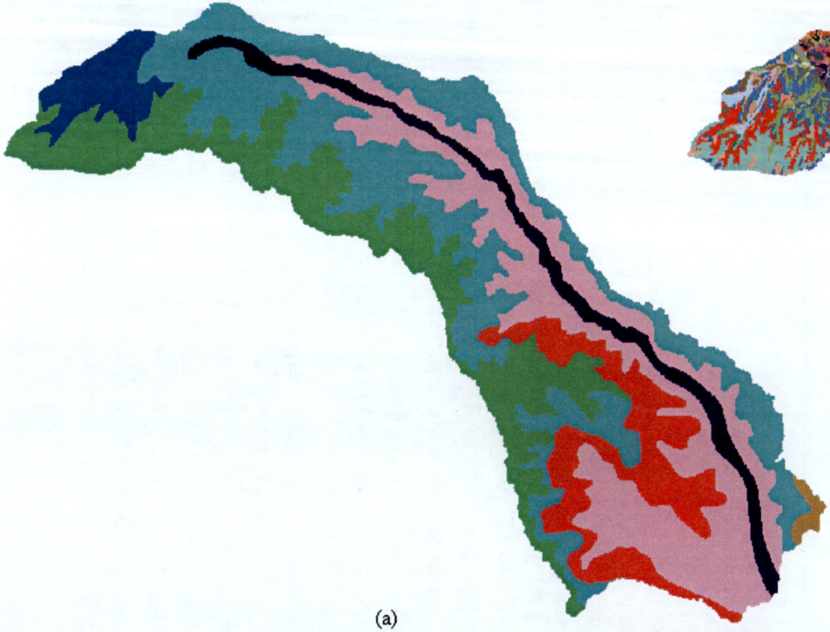

(a)

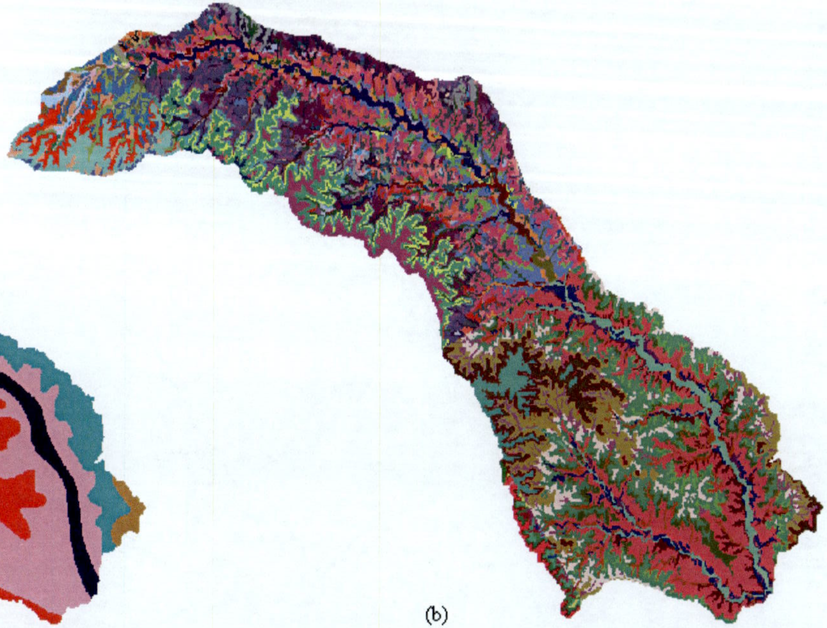

(b)

Figure 4. Map showing classifications of the soils in the study area within the Cowhouse Creek watershed, located in north central Texas, in the two U.S. Department of Agriculture (USDA) Natural Resources Conservation Service (NRCS) soil databases of (a) the State Soil Geographic (STATSGO) with a county-level spatial resolution, and (b) the Soil Survey Geographic (SSURGO) with a farm-level spatial resolution. The purpose is to illustrate that the SSURGO data are much more detailed than the STATSGO data rather than to show the discernable classifications.

unavailable. Further, because of the consistent PET estimation method, the comparisons of the results between these two study areas would be more logical.

The USGS has been monitoring daily streamflows at station USGS 08101000 since 1 October 1950. The data indicated that the Cowhouse Creek was a seasonal stream, with low flows in summer and fall, intermediate flows in winter, and high flows in spring. The daily streamflows varied from 0 to 1000 m^3/s, with an annual average value of 2.8 m^3/s. The streamflows in spring were predominantly generated from rainfall runoff, but the streamflows in other three seasons were predominantly fed by the Edwards-Trinity aquifer (USGS, 2005). The data also indicated that the daily peaks ranged from 0.1 to 996.8 m^3/s and that the instantaneous peaks could be as high as 3115 m^3/s that occurred in 20 December 1991. For a given year, the peak could occur in any season. In this study, the data on daily streamflows for the water years of 1985 through 2006 were used to calibrate and validate the SWAT models, which had a modeling domain covering the upper 62 percent of the Cowhouse Creek watershed.

TXCCW and NDERW have a similar topography, which is characterized by low topographic relief and broad flat alluvial floodplains and river terraces. The dominant soils in these two study areas are clay and sandy clay, with a compatible infiltration rate. However, the land use in NDERW is almost entirely agriculture, whereas, the land use in TXCCW is dominated by range brush and range grasses. The annual average precipitation in TXCCW was approximately twice more than that in NDERW, and the annual mean temperature in TXCCW was about 15°C higher. The streamflows in NDERW were mainly generated from melting snow with negligible base flows, but the streamflows in TXCCW were predominantly generated from rainfall runoff during the high-flow periods and were largely sustained by the aquifer during the low-flow periods. These differences were expected to differentiate these two study areas in terms of the processes of precipitation runoff, soil erosion, and sediment transport (Srinivasan et al., 2005; Conroy et al., 2006).

Model Set up

The USEPA 1:250,000-scale LULC data were downloaded from the USGS website http://edc2.usgs.gov/geodata/index.php and used to set up the models for NDERW, whereas, the 10-m NLCD 2001 data were downloaded from the same website and used for TXCCW. The basic inputs common to the models of these two study areas included the 30-m National Elevation Dataset (NED), National Hydrography Dataset (NHD), STATSGO, and SSURGO (ver. 2.0). NED was developed by merging the highest resolution, best quality elevation data available across the United States into a seamless raster format (USGS, 2001a). NHD is a comprehensive set of digital spatial data that contains information about surface water features such as lakes, ponds, streams, rivers, springs, and wells (USGS, 2001b). This study utilized the NHD stream feature as the reference surface water drainage network to delineate subbasins for modeling purposes. For each of the two study areas, the AvSWAT extension developed by Di Luzio et al. (2002) was used to delineate the boundaries of the entire area and its subbasins, along with their drainage channels. The boundaries for the subbasins were determined by trial and error to ensure that the delineated drainage channels closely match the drainage network presented by the NHD data. As a result, NDERW was subdivided into 53 subbasins, with sizes ranging from 0.4 to 1885 ha, and TXCCW was subdivided into 130 subbasins, with sizes ranging from 5.4 to 4372 ha.

The STATSGO data for the two study areas were downloaded from the USDA-NRCS website http://www.ncgc.nrcs.usda.gov/products/datasets/statsgo. These data required no further preprocessing and were directly taken as the inputs into AvSWAT. The SSURGO data for the three North Dakota counties covering NDERW and for the five Texas counties covering TXCCW were downloaded from the USDA-NRCS website http://www.ncgc.nrcs.usda.gov/products/datasets/ssurgo. The SSURGO data were preprocessed into the AvSWAT required format by implementing a three-step approach for each of the study areas. First, the SSURGO spatial data or coverages for the covering counties were merged into a single coverage. Second, the tabular source data associated with each of the county coverages were separately preprocessed using a computer program written by the authors in the Microsoft Visual Basic® computer language. Upon request, this computer program may be available from the corresponding author. Third, the preprocessed tabular source data were compiled into a user soil database with the AvSWAT required format for the study area. As a result, two single SSURGO coverages and two user soil databases were generated.

For each of the 53 subbasins in NDERW, the LULC and STATSGO data were used to define multiple HRUs for the SWAT-STATSGO model, while the single SSURGO coverage along with the user soil database and LULC were used to define multiple HRUs for the SWAT-SSURGO model. To avoid defining HRUs with a tiny size, the threshold levels of 20% for land use and 10% for soil were used (Wang and Melesse, 2006). The AvSWAT interface defined 108 HRUs for the SWAT-STATSGO model, with 1 to 4 HRUs for a subbasin. For the SWAT-SSURGO model, the interface defined 1 to 5 HRUs for a subbasin, resulting in a total of 135 HRUs for NDERW. The initial values for the parameters used to configure the models were automatically extracted and/or estimated from these datasets by the interface. In SWAT, these parameters are grouped at levels of watershed, subbasin, and HRU and are described in detail by Neitsch et al. (2002a).

Similarly, for each of the 130 subbasins in TXCCW, the NLCD 2001 and STATSGO data were used to define multiple HRUs for the SWAT-STATSGO model, while the single SSURGO coverage along with the user soil database and the NLCD 2001 data were used to define multiple HRUs for the SWAT-SSURGO model. With the threshold levels of 20% for land use and 10% for soil, the AvSWAT interface defined 546 HRUs for the SWAT-STATSGO model, with 1 to 8 HRUs for these subbasins. For the SWAT-SSURGO model, the interface defined 3 to 17 HRUs for these subbasins, resulting in a total of 880 HRUs for TXCCW. Again, the initial values for the parameters used to configure the models were automatically extracted and/or estimated from these datasets by the interface. For description purposes, the two models for NDERW were designated SWAT-STATSGO-NDERW and SWAT-SSURGO-NDERW, whereas, the two models for TXCCW were designated SWAT-STATSGO-TXCCW and SWAT-SSURGO-TXCCW.

For NDERW, the data on daily precipitation and minimum and maximum temperatures of the two NWS stations PR321686 and PR325660 (Figure 1) were preprocessed into database files with the SWAT required format for an evaluation period extending from 1 October 1981 to 30 November 1986. In addition, the available data on daily precipitation and minimum and maximum temperatures for these two NWS stations from 1 October 1975 to 30 September 1981 were also preprocessed and used to define the model initial conditions (i.e., initial values for the model parameters such as soil moisture and depth of water in shallow and deep aquifers). The missing values on precipitation and temperatures for these eight years were simulated by the weather generator that is a component of the SWAT software package (Neitsch et al., 2002b).

For TXCCW, the data on daily precipitation and minimum and maximum temperatures of the NWS station PR413485 (Figure 3) were preprocessed into database files with the SWAT required format for a simulation period extending from 1 January 1980 to 31 December 2006. The period from 1 January 1980 to 30 September 1984 were used to define the initial values of the model parameters. The few missing values on precipitation were simulated by the weather generator.

Assessment Method

The daily streamflows observed at station USGS 05062200 from 1 December 1984 to 30 November 1986 were used to independently calibrate the SWAT-STATSGO-NDERW and SWAT-SSURGO-NDERW models, which then were validated using the observed daily streamflows at the station from 1 December 1981 to 30 November 1984. On the other hand, the daily streamflows observed at station USGS 08101000 from 1 December 1984 to 30 November 1996 were used to independently calibrate the SWAT-STATSGO-TXCCW and SWAT-SSURGO-TXCCW models, which then were validated using the observed daily streamflows at the station from 1 December 1996 to 30 November 2006. The calibrations were implemented by manually adjusting the parameters listed in Table 1 to make the differences as small as possible over the calibration periods between: the model generated and observed daily streamflows; the monthly volumes calculated on the basis of the modeled and observed daily streamflows; and the exceedence times for various flow thresholds calculated on the basis of the modeled and observed daily streamflows (Wang and Melesse, 2005, 2006).

The model performance was judged using visualization plots showing modeled versus observed values and two commonly used statistics, namely coefficient of determination (R^2; Johnson and Wichern, 1998) and Nash-Sutcliffe coefficient (E^2; Nash and Sutcliffe, 1970). E^2 is computed as:

$$E^2 = 1 - \frac{\sum_{i=1}^{n} (Q^i_{obs} - Q^i_{sim})^2}{\sum_{i=1}^{n} (Q^i_{obs} - \overline{Q}_{obs})^2} \qquad (1)$$

where Q^i_{sim} and Q^i_{obs} are the simulated and observed streamflows, respectively, on the ith time step, and $\overline{Q}_{obs}$ is the average of Q^i_{obs} across the n evaluation time steps.

The value of E^2 can range from $-\infty$ to 1.0, with higher values indicating a better overall fit and 1.0 indicating a perfect fit. A negative E^2 indicates that the simulated streamflows are less reliable than using the average of the observed streamflows, while a positive value indicates that they are more reliable than using this average. Based on Motovilov et al. (1999), the simulated streamflows are considered "good" for values of $E^2 > 0.75$, while for values of E^2 between 0.75 and 0.36, the simulated streamflows are considered "satisfactory."

Because observed data on sediment concentrations or loadings were unavailable for the two study areas, the sediment components of the models could not be calibrated and validated. However, the six sediment-related parameters, namely ADJ_PKR, PRF, SPCON, SPEXP, CH_EROD, CH_COV (Table 1), were empirically adjusted to make the predicted annual average sediment yield at station USGS 05062200 comparable with the value reported by Stoner et al. (1993) and the predicted sediment yield at USGS 08101000 comparable with the value reported by Narasimhan et al. (2007). Based on a analysis of the geological settings, Stoner et al. (1993) estimated that the average annual sediment yield in the Red River of the North Basin, in which NDERW is located, would be 0.35 tonnes/ha, a typical value in the glacial prairie region (Ashmore, 1993). In the Cedar Creek watershed, Narasimhan et al. (2007) conducted a comprehensive field survey at 56 sites across the watershed and in one lake. Subsequently, using the Rapid Assessment Point Method (RAP-M) developed by Windhorn (2001) and adopted by NRCS, the authors estimated that the average annual total sediment yield for the Cedar Creek watershed would be 1.72 tonnes/ha, of which 0.59 tonnes/ha could be originated from the stream bank/bed erosion and 1.13 tonnes/ha from the overland erosion. Because the Cowhouse Creek watershed is about only about 150 km south of the Cedar Creek watershed, this study assumed that the sediment yield for TXCCW would have a similar magnitude and proportional pattern (i.e., 34.3% of the sediment from the stream bank/bed erosion and 65.7% from the overland erosion).

For each of the two study areas, the statistics were computed and the duration curves plotted from the observed and model predicted streamflows at the outlet for both the calibration and validation periods. The SWAT-STATSGO model was judged to be better than the SWAT-SSURGO model when its corresponding values for R^2 and/or E^2 were greater. Otherwise, the SWAT-SSURGO model was judged to be better when its corresponding values for R^2 and/or E^2 were larger. Whichever model resulted in a duration curve closer to the duration curve of the observed streamflows was judged to have a better performance on predicting the range in magnitude of the flows. In addition, the sediment loadings at the outlet predicted by the two models were compared to assess the effects of using one soil database over another.

Further, for each subbasin, the predicted daily sediment yields by a model were arithmetically averaged across the entire evaluation period to calculate the corresponding means. Also, for each subbasin, the frequency associated with the predicted sediment yields was calculated as the ratio of the number of days with a nonzero sediment contribution to the total number of evaluation days (i.e., 8035 for TXCCW and 120 for NDERW). The calculated means and frequencies were mapped as low, medium, and high categories, and visually compared. For each study area, the three categories were separately defined based on the first and third quartiles calculated using the pooled datasets of the corresponding means. For example, for the map of mean sediment yield, the subbasin means predicted by the two models were pooled together to generate a dataset. This dataset was used to calculate the first and third quartiles. Rounded to the nearest zero or fifth hundredth, the two quartiles were taken as the broken points of the three categories. For this purpose, the low category was defined as having a value less than the first quartile, the high category, on the other hand, was defined as having a value greater than the third quartile. The medium category was defined as having a value between these two quartiles.

When the difference between the sediment yields predicted by the two models was larger for TXCCW than that for NDERW, the soil data resolution was judged to be more sensitive to the simulation of sediment for TXCCW. Otherwise, the soil data resolution was judged to be more sensitive to the simulation of sediment for NDERW.

Results and Discussion

Calibrated Models

For the SWAT-STATSGO-NDERW model, the default (i.e., AvSWAT generated) values for CN2 were reduced by 8, whereas, for the SWAT-SSURGO-NDERW model, the values were adjusted down by 7 (Table 1). Because SWAT uses CN2 to control the estimation of excess precipitation (Neitsch et al., 2002b; USDA-NRCS, 2004), this indicated that using STATSGO over SSURGO would be sensitive to the simulation of runoff processes in the NDERW study area. However, these two models took the default values for REVAPMN and GW_REVAP. This indicated that AvSWAT can reasonably estimate these two groundwater evapotranspiration related parameters, while for a given subbasin, the two models had distinctly different values for each of these two parameters. In addition, the two models took an identical value for each of the other 15 adjusted parameters (Table 1).

On the other hand, the SWAT-STATSGO-TXCCW and SWAT-STATSGO-TXCCW models took the AvSWAT estimated values for CN2 in terms of the soils presented by the STATSGO and SSURGO data, respectively. This indicated that using STATSGO over SSURGO would not be as sensitive as in the NDERW study area to the simulation of runoff process in the TXCCW study area. In contrast to the two North Dakota models, the two Texas models took distinctly different values for ESCO, CH_COV, and CH_EROD, respectively (Table 1). SWAT uses ESCO to control the estimation of soil moisture evaporation (Neitsch et al., 2002b), and CH_COV and CH_EROD to control the prediction of stream bank/bed erosion and sediment deposition. Thus, for the TXCCW study area, using STATSGO over SSURGO would be sensitive to the simulation of evapotranspiration process and sediment transport. In addition, the three snowmelt related parameters, namely SMTMP, SMFMX, and TIMP, were very sensitive for the two North Dakota models (Wang and Melesse, 2005), but as mentioned above, these three parameters had no influence on the simulation of the TXCCW study area. Further, the two Texas models were determined to have an identical value for each of the other 10 adjusted parameters (Table 1).

The four models were calibrated to have a marginally good or good performance on simulating the monthly and seasonal mean discharges at the outlets of the corresponding study areas ($E^2 > 0.60$; Table 2). In terms of the observed streamflow volumes, the two North Dakota models had a prediction error of less than 10%, whereas, the two Texas models had a prediction error of about only 1%. The discrepant prediction accuracies further indicated the sensitivity of using STATSGO over SSURGO to the simulation of runoff processes in the NDERW study area.

Table 1. List of adjusted parameters.

Parameter	Definition	Range[a]	NDERW[c]		TXCCW[d]	
			SWAT-STATSGO[e]	SWAT-SSURGO[f]	SWAT-STATSGO[e]	SWAT-SSURGO[f]
SMTMP (°C)	Snowmelt base temperature	0.0 – 3.0	1.5	1.5	No Influence	No Influence
SMFMX (mm H_2O/°C–day)	Maximum snowmelt factor	1.4 – 6.9	6.9	6.9	No Influence	No Influence
TIMP	Snowpack temperature lag factor	0.01 – 1.0	0.3	0.3	No Influence	No Influence
SURLAG (day)	Surface runoff lag coefficient	1.0 – 12.0	1.0	1.0	2.0	2.0
MSK_CO1	Muskingum translation coefficient for normal flow	0.0 – 10.0	1.2	1.2	2.5	2.5
MSK_CO2	Muskingum translation coefficient for low flow	0.0 – 10.0	1.2	1.2	2.5	2.5
REVAPMN (mm H_2O)	Threshold depth of water in the shallow aquifer for "revap" or percolation to the deep aquifer to occur	0.0 – 500.0	Default[b]	Default[b]	1.0	1.0
GW_REVAP	Groundwater "revap" coefficient	0.0 – 0.2	Default[b]	Default[b]	0.15	0.15
GWQMN (mm H_2O)	Threshold depth of water in the shallow aquifer required for return flow to occur	0.0 – 5000.0	500	500	500	500
ESCO	Soil evaporation compensation factor	0.01 – 1.0	0.95	0.95	0.27	0.30
EPCO	Plant water uptake compensation factor	0.01 – 1.0	0.95	0.95	0.85	0.85
CN2	SCS curve number for soil moisture condition II	60.0 – 95.0	Reduce by 8	Reduce by 7	Default[b]	Default[b]
CH_COV	Channel cover factor	0.0 – 1.0	0.25	0.25	0.022	0.500
CH_EROD (cm/hr/pa)	Channel erodibility factor	0.0 – 1.0	0.15	0.15	0.090	0.150
ADJ_PKR	Peak rate adjustment factor for sediment routing in subbasin	1.0 – 3.0	1.20	1.20	1.52	1.50
PRF	Peak rate adjustment factor for sediment routing in reach	1.0 – 3.0	1.00	1.00	1.00	1.00
SPCON	Coefficient in sediment transport equation	0.0001 – 0.01	0.0001	0.0001	0.05	0.05
SPEXP	Exponent in sediment transport equation	1.0 – 2.0	1.00	1.00	1.50	1.50

[a] The ranges for the parameters were based on Neitsch et al. (2002a, b) and Wang and Melesse (2005, 2006).

[b] The values for this parameter was derived by the AvSWAT interface and varied from one hydrologic response unit (HRU) to another.

[c] The study area located within the Elm River watershed, North Dakota.

[d] The study area located within the Cowhouse Creek watershed, Texas.

[e] The SWAT model with the State Soil Geographic (STSTSGO) database as an input.

[f] The SWAT model with the Soil Survey Geographic (SSURGO) database as an input.

Table 2. Statistics of daily, monthly, and seasonal discharges for the evaluation water years (December through November) of the study area in the Elm River watershed in eastern North Dakota.[a]

Statistics	Calibration[b]							Validation[c]						
	Obs. (m^3/s)	SWAT–STATSGO			SWAT–SSURGO			Obs. (m^3/s)	SWAT–STATSGO			SWAT–SSURGO		
		Pred. (m^3/s)	R^2	E^2	Pred. (m^3/s)	R^2	E^2		Pred. (m^3/s)	R^2	E^2	Pred. (m^3/s)	R^2	E^2
Daily	0.33	0.36	0.53	0.51	0.32	0.51	0.49	0.72	0.74	0.55	0.31	0.70	0.55	0.26
Monthly	0.33	0.36	0.89	0.88	0.31	0.92	0.92	0.72	0.74	0.53	0.50	0.70	0.53	0.49
Seasonal	0.33	0.36	0.98	0.97	0.31	0.97	0.97	0.71	0.74	0.69	0.67	0.70	0.67	0.63

[a] Obs. is the observed value, Pred. is the predicted value, R^2 is the coefficient of determination, and E^2 is the Nash-Sutcliffe coefficient defined by Equation (1)
[b] The calibration period is from 1 December 1984 to 30 November 1986.
[c] The validation period is from 1 December 1981 to 30 November 1984.

Table 3. Statistics of daily, monthly, and seasonal discharges for the evaluation water years (December through November) of the study area in the Cowhouse Creek watershed in north central Texas.[a]

Statistics	Calibration[b]							Validation[c]						
	Obs. (m^3/s)	SWAT–STATSGO			SWAT–SSURGO			Obs. (m^3/s)	SWAT–STATSGO			SWAT–SSURGO		
		Pred. (m^3/s)	R^2	E^2	Pred. (m^3/s)	R^2	E^2		Pred. (m^3/s)	R^2	E^2	Pred. (m^3/s)	R^2	E^2
Daily	3.02	3.04	0.05	-0.11	2.99	0.05	-0.10	3.65	3.96	0.03	-0.47	3.84	0.03	-0.46
Monthly	3.02	3.04	0.64	0.63	2.99	0.61	0.60	3.65	3.96	0.22	-0.07	3.84	0.22	-0.05
Seasonal	3.02	3.04	0.79	0.76	2.99	0.75	0.72	3.65	3.96	0.04	-0.37	3.84	0.04	-0.37

[a] Obs. is the observed value, Pred. is the predicted value, R^2 is the coefficient of determination, and E^2 is the Nash-Sutcliffe coefficient defined by Equation (1)
[b] The calibration period is from 1 December 1984 to 30 November 1996.
[c] The validation period is from 1 December 1996 to 30 November 2006.

The good model performance was also verified by examining the visualization plots showing the observed versus predicted streamflows (Figures 5 and 10). For the NDERW study area, the annual average discharges predicted by the two models were very close to the corresponding observed values (Figure 5a). Compared with the SWAT-SSURGO-NDERW model, the SWAT-STATSGO-NDERW model tended to better predict the monthly mean peak discharge in March, but tended to overestimate the discharges for the recession months from May to July (Figure 5b). These two models had a comparable performance on simulating the discharges for the other months.

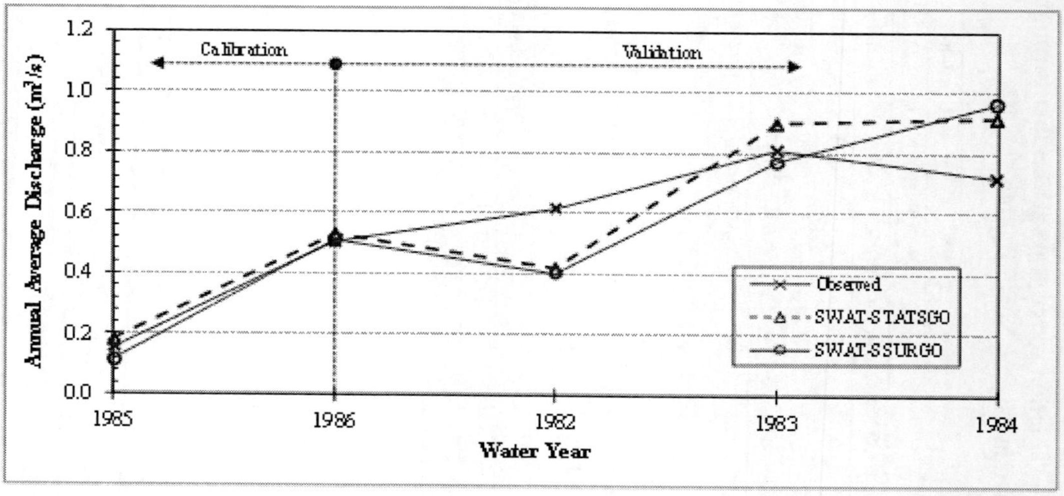

(a)

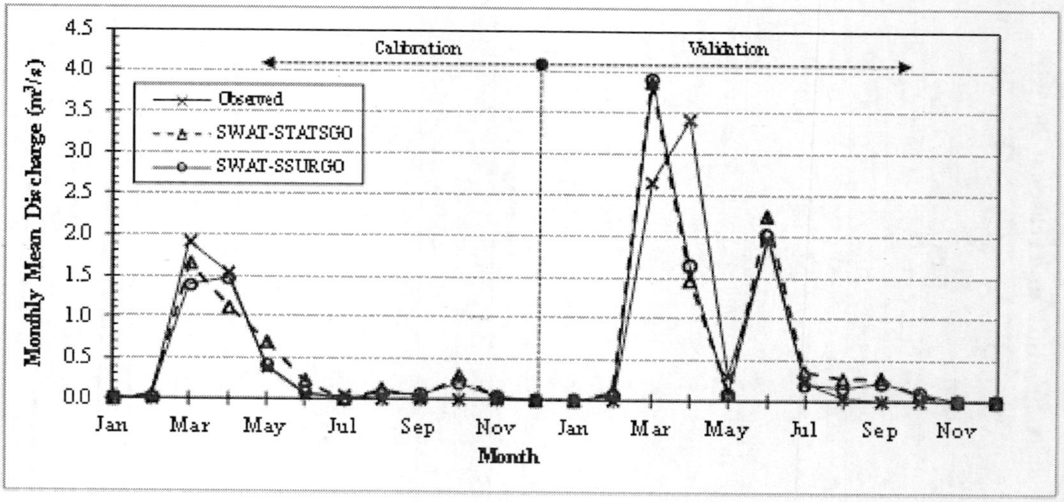

(b)

Figure 5. Plots showing the observed versus model predicted (a) annual average and (b) monthly mean discharges at the outlet of NDERW, the study area within the Elm River watershed in eastern North Dakota.

For the TXCCW study area, the two models overestimated the annual average discharges for some water years, while for the other years, the discharges were underpredicted (Figure 10a). In particular, the models noticeably underestimated the streamflow occurred in 1991, when a record-breaking peak discharge was recorded at USGS 08101000 on December 20 as a result of an extreme storm (USGS, 1998). The storm had daily rainfall totals exceeded 102 mm at numerous locations across the TXCCW study area. This was because SWAT may be not powerful in simulating large extreme storm events (Wang and Melesse, 2006) such as the

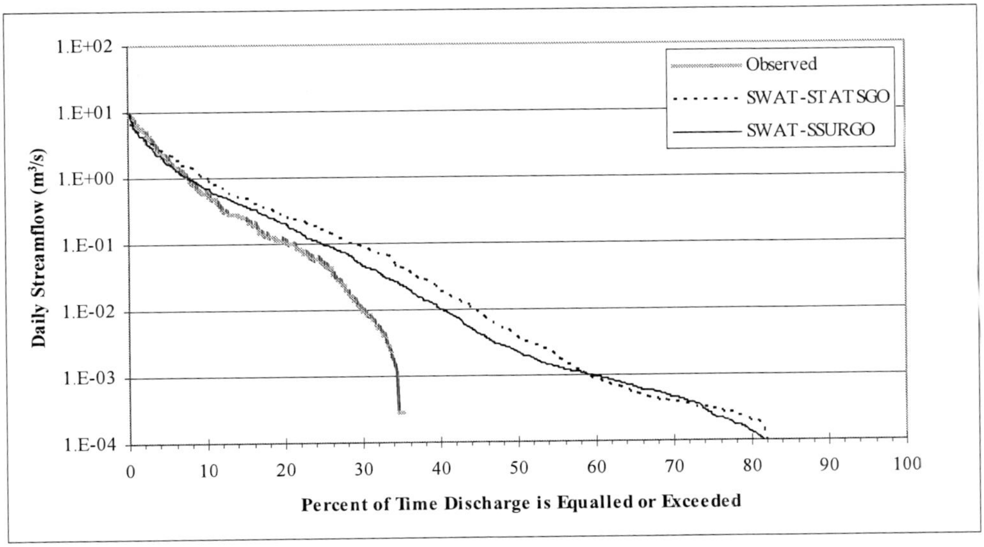

(a)

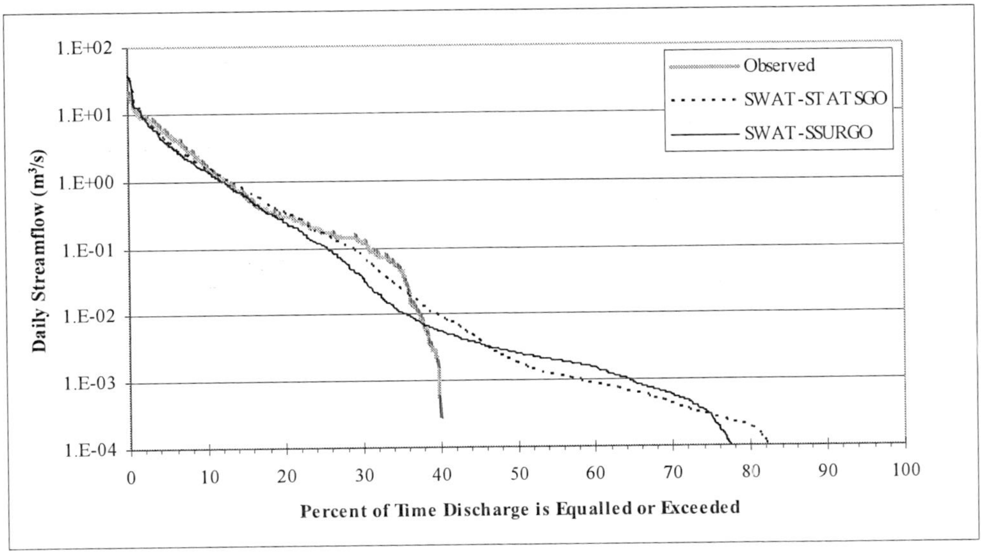

(b)

Figure 6. Duration curves of the observed and model predicted daily streamflows at the outlet of NDERW, the study area within the Elm River water in eastern North Dakota for the (a) calibration period and (b) validation period.

one occurred in 1991. During large events, the soil moisture conditions could have a very large variation within a day. However, with a daily simulation time step, the SCS curve number method tends to use a daily average soil moisture condition, resulting in the underestimation of the high streamflows. In contrast with the two North Dakota models, the two Texas models had a comparable performance on simulating the monthly mean discharges and the predicted discharges were very close to the corresponding observed values (Figure 10b). The small simulation discrepancies for monthly mean discharges indicated that using STATSGO over SSURGO would not be very sensitive to the simulated trend of streamflows, whereas, the under and/or overestimation of the annual average discharges indicated that using STATSGO over SSURGO could noticeably affect the predicted volumes for this Texas study area.

The North Dakota models had a satisfactory performance on simulating the daily streamflows ($E^2 > 0.49$), the Texas models, however, had a poor performance ($E^2 < 0$). The SWAT-STATSGO-NDERW and SWAT-SSURGO-NDERW models did a comparable job on simulating streamflows higher than 4 m^3/s but tended to overpredict lower flows (Figure 6a). The overprediction might be because the large errors of the observed values (Wang and Melesse, 2006). In contrast, the SWAT-STATSGO-TXCCW and SWAT-SSURGO-TXCCW models underestimated streamflows higher than 100 m^3/s and lower than 50 m^3/s, but the models overestimated the flows between these two values (Figure 11a). Again, this might be due to the SWAT's inability to simulate short-term thunderstorms as well as very dry conditions (Feyereisen et al., 2007). For this reason, the predicted sediment loadings and yields were evaluated at monthly and annual time steps only.

For the TXCCW study area and during the validation period, the two models were unable to successfully simulate another two extreme large storms occurred in 1997 and 2004 (Figure 10), resulting in negative E^2 values (Table 2). Nevertheless, the two models did a good job on estimating the volumes of the observed streamflows, as indicated by a low prediction error of less than 8%. For the NDERW study area and during the validation period, because the hydrologic conditions were similar with those during the calibration period (Wang and Melesse, 2006), the two models had a satisfactory performance on simulating the monthly and seasonal mean discharges ($E^2 > 0.49$). The simulated daily discharges were more reliable than using the average of the observed daily streamflows ($E^2 > 0.26$). Overall, the four models had a comparable performance with those reported by other researchers (e.g., Vazquez-Amábile et al., 2005; Du et al., 2006; Van Liew et al., 2007). Using these four calibrated models, this study evaluated influences of using STATSGO over SSURGO on the prediction of sediment source areas within NDERW and TXCCW for the calibration as well as validation periods as a whole, i.e., for the entire evaluation periods.

Predicted Sediment

The two North Dakota models were determined to take an identical value for each of the six sediment related parameters, whereas, the two Texas models had different values for CH_COV, CH_EROD, or ADJ_PKR but had an identical value for PRF, SPCON, or SPEXP (Table 1). Again, this indicated that using STATSGO over SSURGO would be more sensitive to the simulation of sediment in the TXCCW study area than in the NDERW study area. With the adjusted values of these six parameters, the annual average sediment yield at the NDERW outlet predicted by the SWAT-STATSGO-NDERW model was almost identical to that predicted by the SWAT-SSURGO-NDERW model (0.38 versus 0.37 tonnes/ha). In addition, the predicted sediment yields were judged to be reasonably close to the regional typical value of 0.35 tonnes/ha reported by Ashmore (1993).

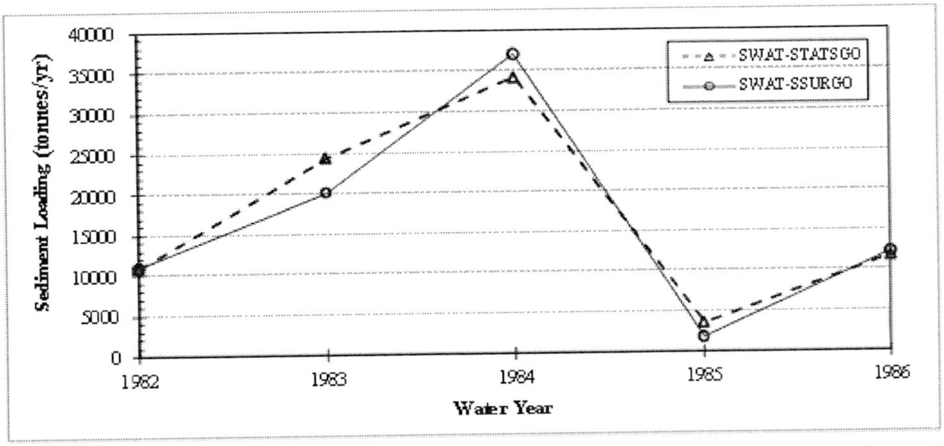

Figure 7. Plot showing the model predicted sediment loadings at the outlet of NDERW, the study area within the Elm River watershed in eastern North Dakota.

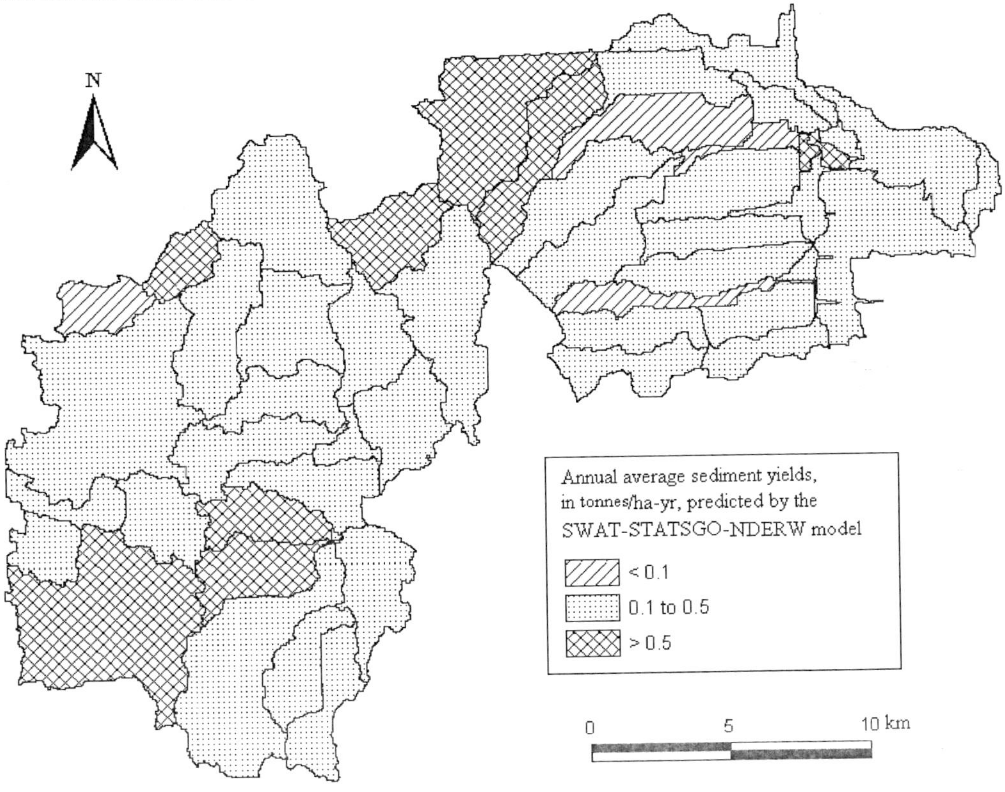

(a)

Figure 8. Continued on next page.

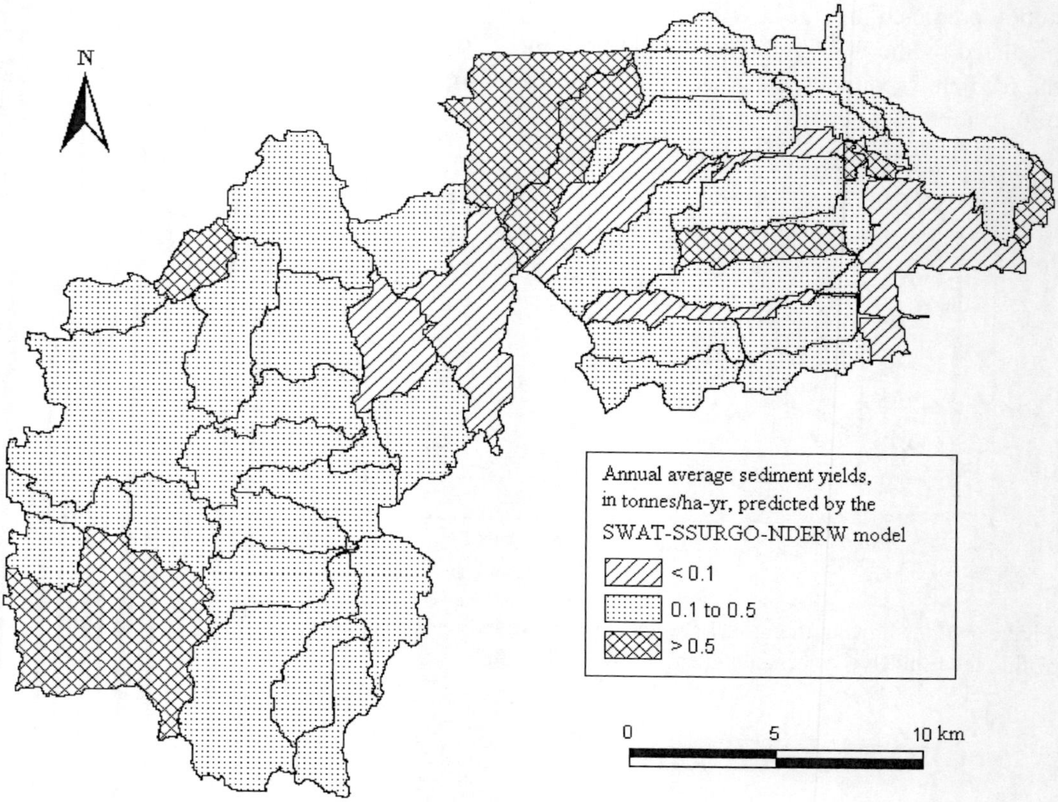

(b)

Figure 8. Maps showing the (a) SWAT-STATSGO-NDERW and (b) SWAT-SSURGO-NDERW predicted sediment yields from the subbasins of NDERW, the study area within the Elm River watershed in eastern North Dakota.

The SWAT-STATSGO-TXCCW model was determined to have a larger value for ADJ_PKR, indicating that for the TXCCW study area, using the STATSGO data instead of the SSURGO data tended to predict more sediment deposition. Conversely, the SWAT-SSURGO-TXCCW model was determined to have larger values for CH_COV and CH_EROD, indicating that using the STATSGO data instead of the SSURGO data tended to predict more stream bank/bed erosion. With the adjusted values for the six sediment related parameters (Table 1), the SWAT-STATSGO-TXCCW model predicted that the annual average sediment yield at the TXCCW outlet would be 1.76 tonnes/ha, of which 0.66 tonnes/ha would be originated from the stream bank/bed erosion and 1.10 tonnes/ha from the overland erosion, whereas, the SWAT-SSURGO-TXCCW model predicted that the sediment yield would be 1.79 tonnes/ha, of which 0.65 tonnes/ha would be originated from the stream bank/bed erosion and 1.14 tonnes/ha from the overland erosion. Obviously, the sediment yields predicted by the two models were almost identical and the predicted values were judged to be reasonably comparable with the results reported by Narasimhan et al. (2007).

Compared with the SWAT-SSURGO-NDERW model, the SWAT-STATSGO-NDERW mode tended to give a higher sediment yield at the NDERW outlet for the evaluation years except for 1984 (Figure 7).This was consistent with the trend of the predicted annual average discharges (Figure 5a) because a higher discharge would have more energy to erode soils and/or stream banks/beds, and to transport the subsequent sediments out of the study area (Williams, 1995). Another explanation might be that the SWAT-SSURGO-NDERW model tended to predict more base flows as inferred by Peschel et al. (2006). In addition, the SWAT-STATSGO-NDERW model predicted that more subbasins within the NDERW study area would have medium (0.1 to 0.5 tonnes/ha) and/or high (> 0.5 tonnes/ha) annual average sediment yields (Figure 8a versus 8b). However, a subbasin that was predicted by the SWAT-STATSGO-NDERW model to more frequently contribute sediment, i.e., to have fewer days with a zero sediment yield, might be predicted by the SWAT-SSURGO-NDERW model to least frequently contribute sediment, and vice versa (Figure 9a versus 9b). Therefore, while the two models predicted a similar sediment yield at the NDERW outlet, they would give distinctly different predictions of sediment contribution sources within the study area.

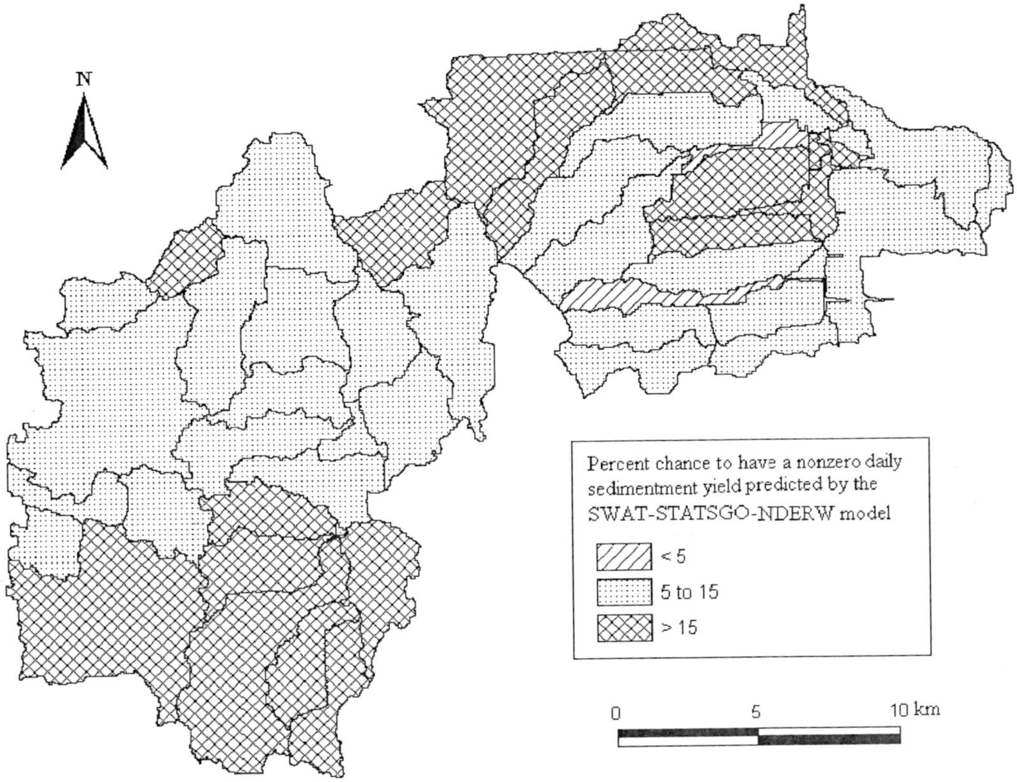

(a)

Figure 9. Continued on next page.

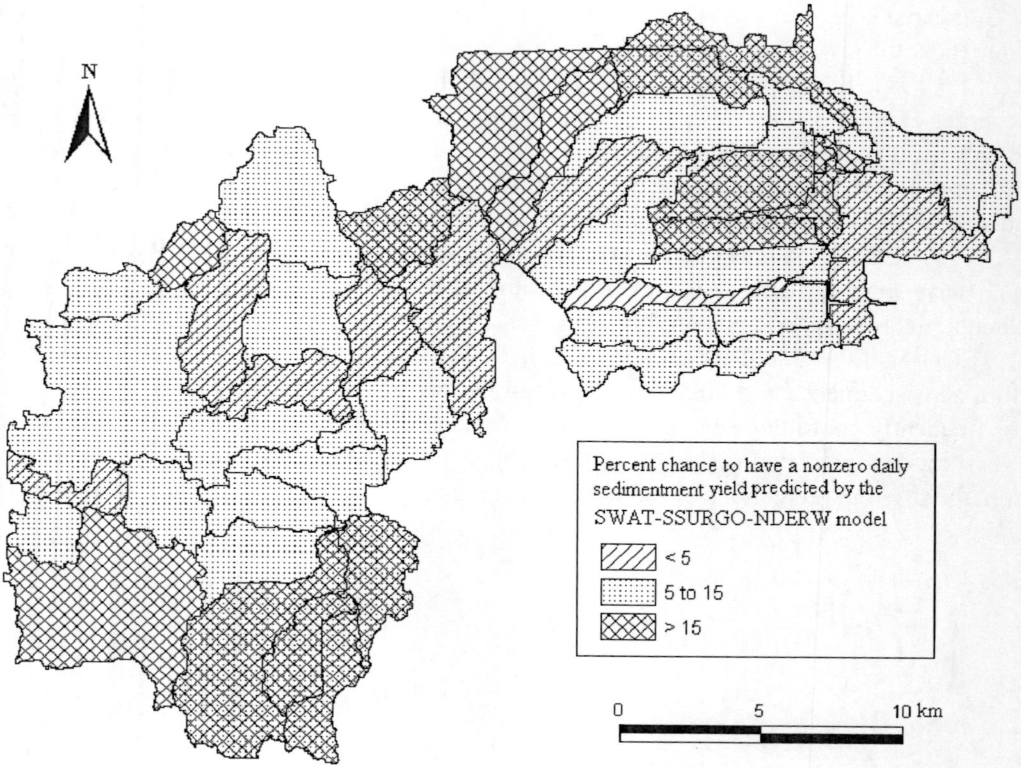

(b)

Figure 9. Maps showing the (a) SWAT-STATSGO-NDERW and (b) SWAT-SSURGO-NDERW predicted percent chances to have nonzero sediment yields from the subbasins of NDERW, the study area within the Elm River watershed in eastern North Dakota.

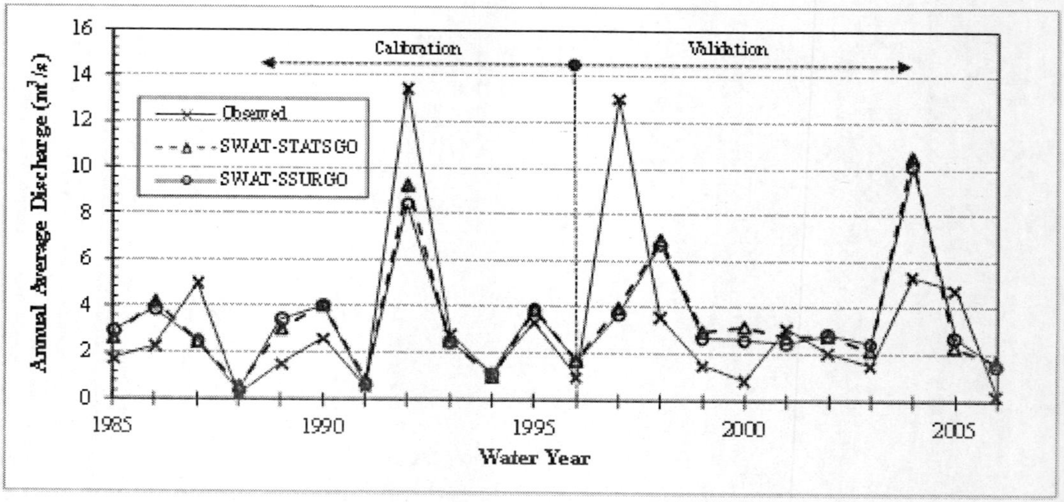

(a)

Figure 10. Continued on next page.

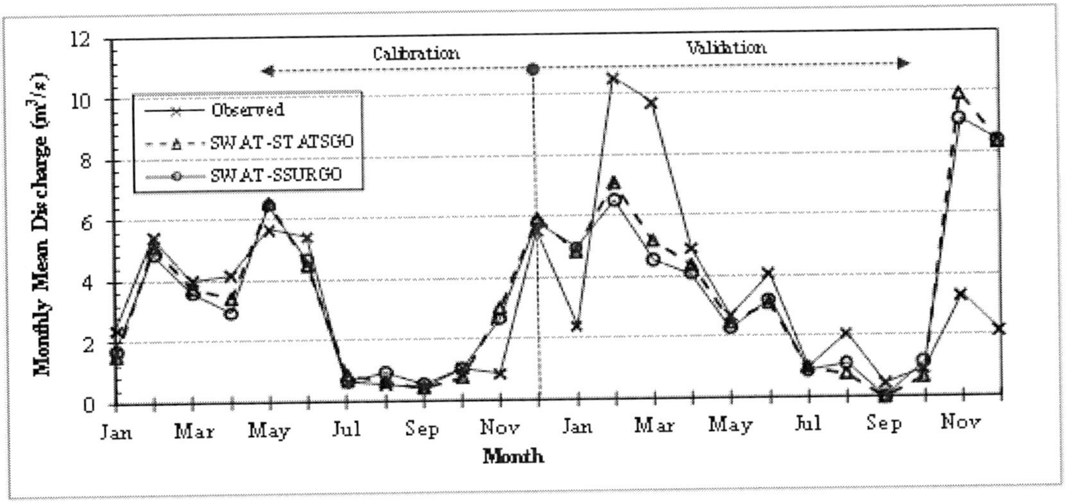

(b)

Figure 10. Plots showing the observed versus model predicted (a) annual average and (b) monthly mean discharges at the outlet of TXCCW, the study area within the Cowhouse Creek watershed in north central Texas.

In contrast, at the TXCCW outlet, the sediment yields predicted by the SWAT-STATSGO-TXCCW model were higher than those predicted by the SWAT-SSURGO-TXCCW model for some years but lower for the other years (Figure 12). Again, this was consistent with the trend of the predicted annual average discharges (Figure 10a). For the upper portion of the study area, the SWAT-STASGO-TXCCW model predicted that more subbasins would have medium (5 to 10 tonnes/ha) and/or high (> 10 tonnes/ha) sediment yields, whereas, for the lower portion of the study area, the SWAT-SSURGO-TXCCW model

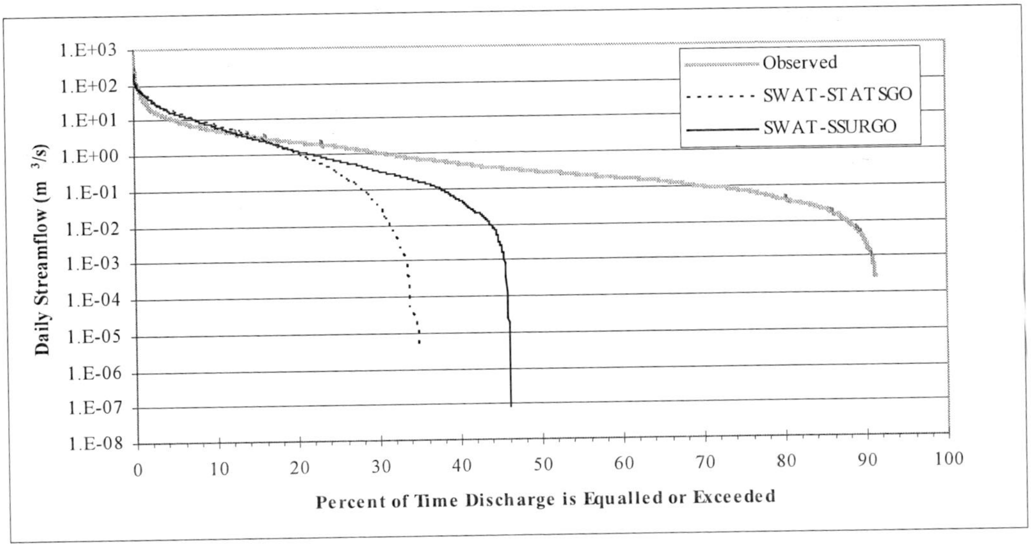

(a)

Figure 11. Continued on next page.

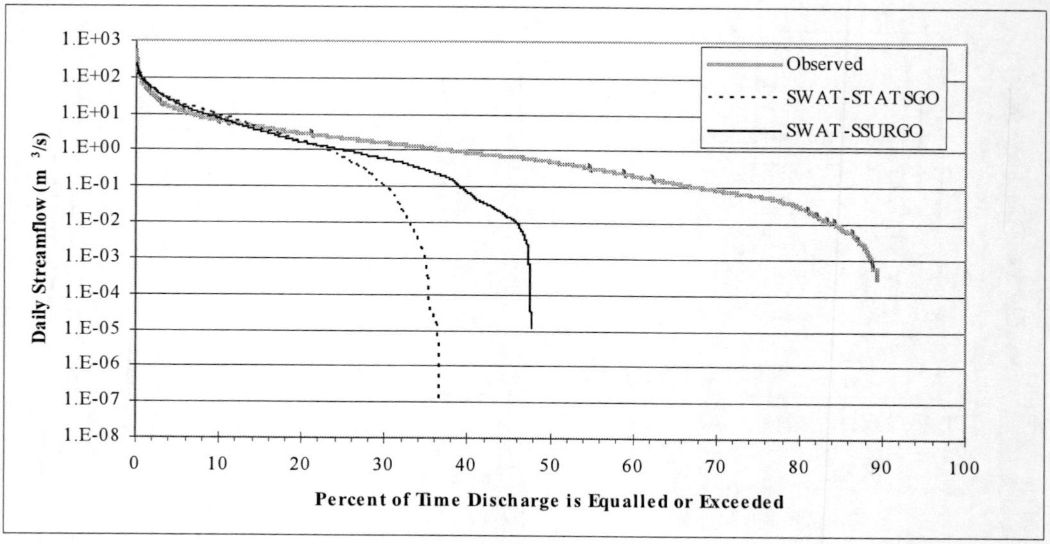

(b)

Figure 11. Duration curves of the observed and model predicted daily streamflows observed at the outlet of TXCCW, the study area within the Cowhouse Creek watershed in north central Texas for the (a) calibration period and (b) validation period.

predicted that more subbasins would have medium and/or high sediment yields (Figure 13). As with the NDERW study area, a subbasin that was predicted by the SWAT-STATSGO-TXCCW model to have a higher sediment contribution frequency might be predicted by the SWAT-SSURGO-TXCCW model to less frequently contribute sediment, and vice versa (Figure 14a versus 14b). Thus, while the two models predicted a similar sediment yield at the TXCCW outlet, they would give distinctly different predictions of sediment contribution sources within the study area.

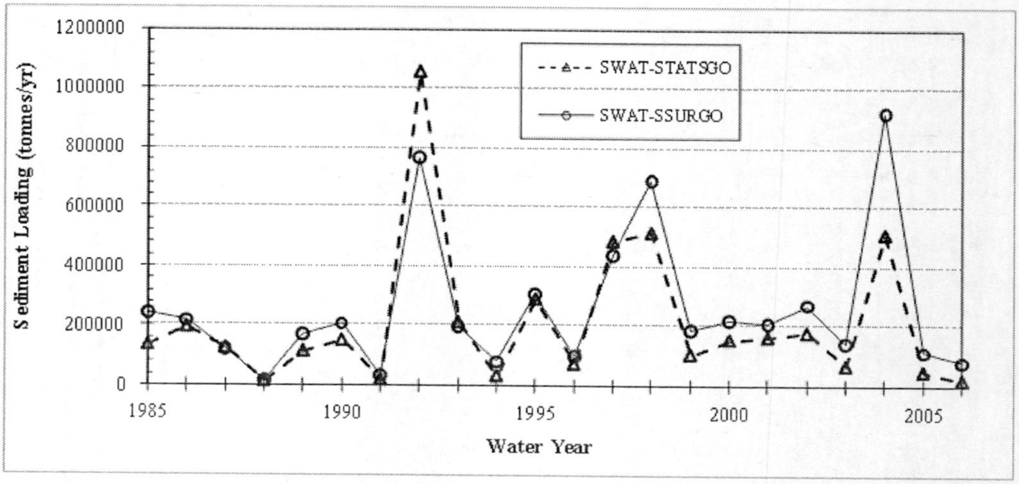

Figure 12. Plot showing the model predicted sediment loadings at the outlet of TXCCW, the study area within the Cowhouse Creek watershed in north central Texas.

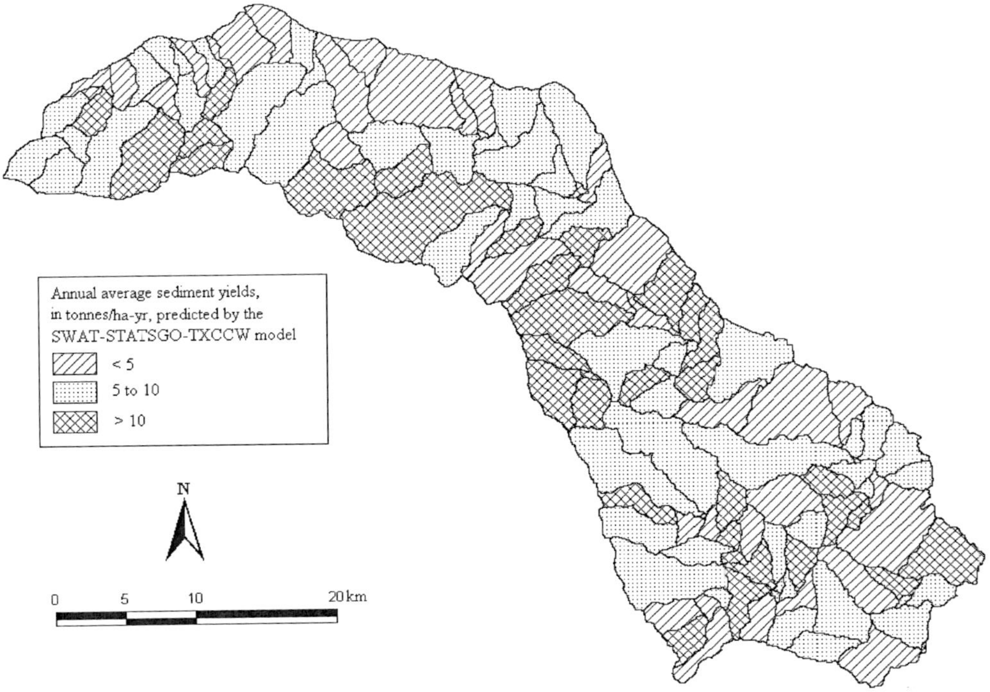

(a)

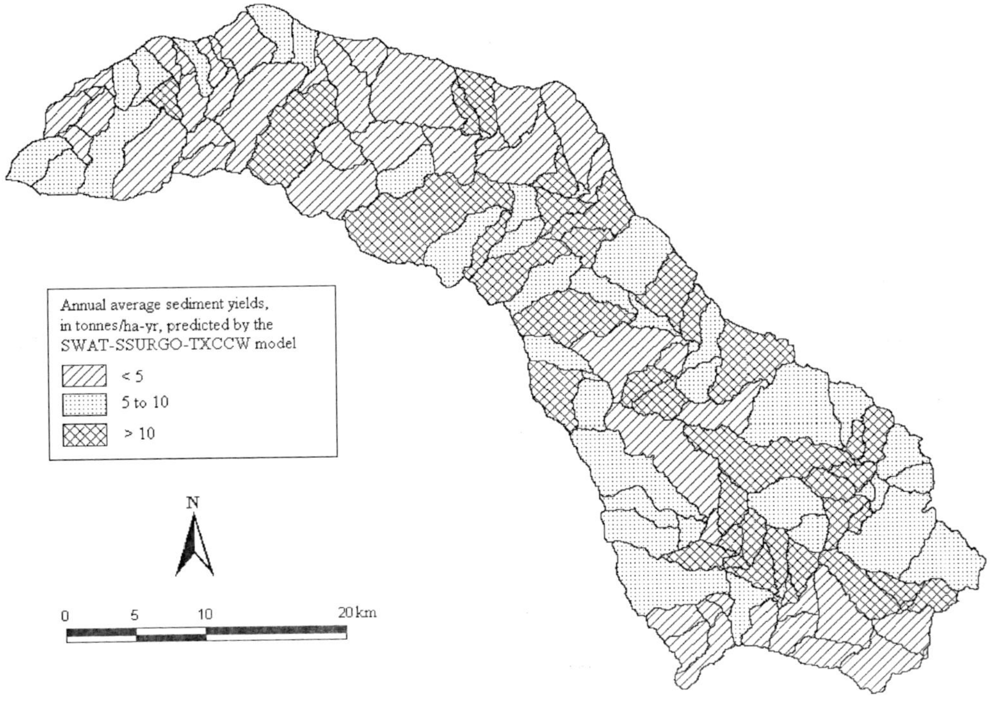

(b)

Figure 13. Maps showing the (a) SWAT-STATSGO-TXCCW and (b) SWAT-SSURGO-TXCCW predicted sediment yields from the subbasins of TXCCW, the study area within the Cowhouse Creek watershed in north central Texas.

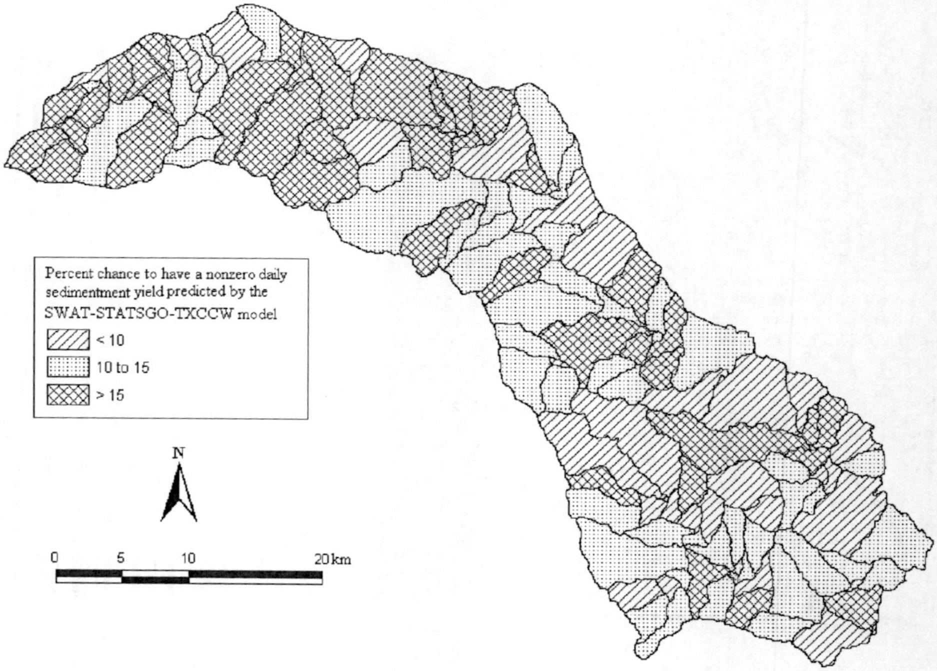

(a)

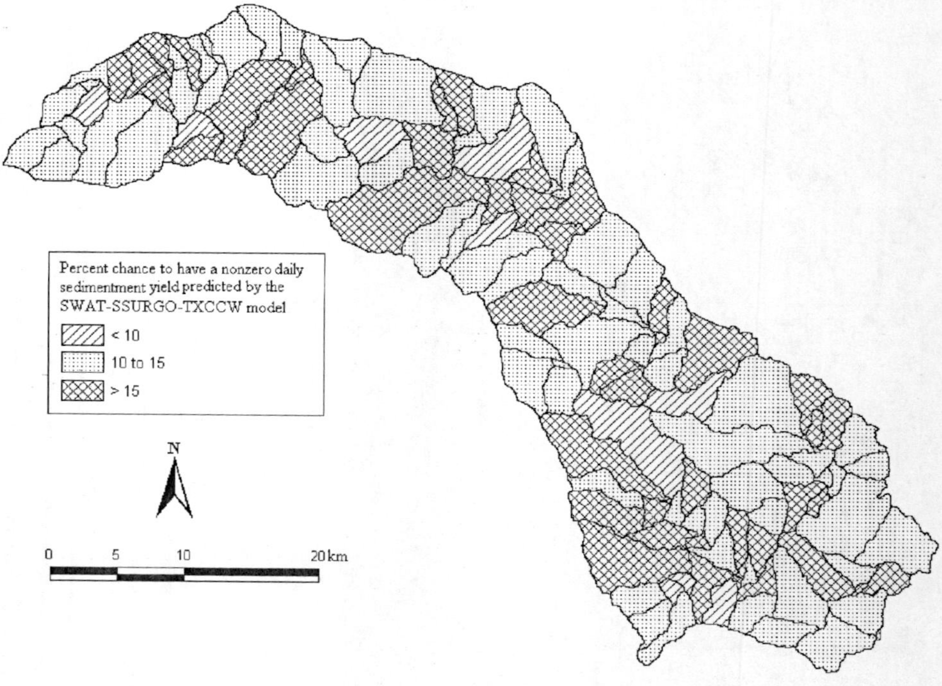

(b)

Figure 14. Maps showing the (a) SWAT-STATSGO-TXCCW and (b) SWAT-SSURGO-TXCCW predicted percent chances to have nonzero sediment yields from the subbasins of TXCCW, the study area within the Cowhouse Creek watershed in north central Texas.

Conclusions

This study evaluated the affects of using STATSGO versus SSURGO on the SWAT's simulation of sediment source areas within the NDERW and TXCCW study areas. These two study areas had distinctly different climatic and hydrologic conditions because the streamflows in NDERW were mainly generated from snowmelt runoff but the streamflows in TXCCW were predominantly generated from rainfall runoff. For a given year, the peak discharges in NDERW almost always occurred in spring, whereas, the peak discharges in TXCCW could occur in any season. In addition, NDERW consists of glaciated soils (Stoner et al., 1993), but TXCCW consists of alluvial soils (Mace et al., 1999). Usually, alluvial soils are more vulnerable for erosion than glaciated soils (Wright, 2001). Further, the land use within NDERW is almost entirely agriculture, the land use within TXCCW, however, is dominated by range brush and range grasses.

The evaluation was conducted using the two SWAT models for NDERW and the other two SWAT models for TXCCW. For each study area, one model took the STATSGO data as an input, whereas, another model took the SSURGO data as an input. The two models were calibrated to have a comparable performance in terms of predicting the observed daily streamflows at the outlet of the study area. The sediment related parameters were empirically adjusted. Subsequently, the calibrated models were used to simulate the sediment yields from the subbasins within the study area for the entire evaluation period. The results indicated that during the calibration periods, all four models did a marginally good or good job in simulating the monthly and seasonal mean discharges at the outlets, during the validation periods, however, the two North Dakota models had a satisfactory performance but the two Texas models had a poor performance. Compared with the two North Dakota models, the two Texas models were determined to have distinctly different values for more parameters. This indicated that using STATSGO over SSURGO would be more sensitive for TXCCW than NDERW. In addition, for each of the study areas, while the two models had a similar performance in terms of predicting the sediment yields at the outlet, they could give distinctly different predictions of the sediment yields (i.e., sediment contribution amounts and frequencies) of the subbasins. The prediction discrepancies were found to be larger between the two Texas models than those between the two North Dakota models.

Acknowledgement

This study was partially supported by Tarleton State University Organized Research Grant (ORG) under contract 153118.

References

Anderson, E. A, 1976. *A Point Energy and Mass Balance Model for Snow Cover*. NOAA Technical Report NWS 19. Washington, D.C.: U.S. Department of Commerce, National Weather Service.

Arnold, J.G., P.M. Allen, and G. Bernhardt. 1993. A comprehensive surface-groundwater flow model. *Journal of Hydrology* **142**:47–69.

Arnold, J.G., R. Srinivasan, R.S. Muttiah, and J.R. Williams. 1998. Large-area hydrologic modeling and assessment: Part I. Model development. *Journal of the American Water Resources Association* **34**(1):73–89.

Ashmore, P. 1993. Contemporary erosion of the Canadian landscape. *Progress in Physical Geography* **17**(2): 190 – 204.

Bagnold, R.A. 1977. Bedload transport in natural rivers. *Water Resources Research* **3:** 303 – 312.

Chow, V. T., D. R. Maidment, and L. W. Mays, 1988. *Applied Hydrology*. New York, N.Y.: McGraw-Hill.

Chu, T.W., A. Shirmohammadi, H. Montas, and A. Sadeghi, 2004. Evaluation of the SWAT Model's sediment and nutrient components in the piedmont physiographic region of Maryland. *Transactions of the ASAE* **47**(5): 1523 – 1538.

Condie, S.A., and I.T. Webster, 1997. The influence of wind stress, temperature, and humidity gradients on evaporation from reservoirs. *Water Resources Research* **33**(12): 2813 – 2822.

Conroy, W.J., R.H. Hotchkiss, and W.J. Elliot. 2006. A coupled upland-erosion and instream hydrodynamic-sediment transport model for evaluating sediment transport in forested watersheds. *Transactions of the ASABE* **49**(6): 1713 – 1722.

Di Luzio, M., R. Srinivasan, J.G. Arnold, and S.L. Neitsch, 2002. *ArcView Interface for SWAT2000 – User's Guide*. Temple, Texas: Grassland, Soil and Water Research Laboratory, Agricultural Research Service; Blackland Research Center, Texas Agricultural Experiment Station.

Du, B., A. Saleh, D.B. Jaynes, and J.G. Arnold. 2006. Evaluation of SWAT in simulating nitrate nitrogen and atrazine fates in a watershed with tiles and potholes. *Transactions of the ASABE* **49**(4): 949 – 959.

Dunne, T., T.R. Moore, and C.H. Taylor. 1975. Recognition and prediction of runoff-producing zones in humid regions. *Hydrological Sciences Bulletin* **20**:305–327.

Feyereisen, G.W., T.C. Strickland, D.D. Bosch, and D.G. Sullivan. 2007. Evaluation of SWAT manual calibration and input parameter sensitivity in the Little River watershed. *Transactions of the ASABE* **50**(3): 843 – 855.

Fontaine, T. A., T. S. Cruickshank, J. G. Arnold, and R. H. Hotchkiss, 2002. Development of a snowfall-snowmelt routine for mountainous terrain for the Soil Water Assessment Tool (SWAT). *Journal of Hydrology* **262**: 209-223.

Greer, C.H. 2005. *Hydrologic Impacts of Mechanical Shearing of Ashe Juniper in Coryell County, Texas*. College Station, Texas: M.S. thesis.

Gupta, H.V., S. Sorooshian, and P.O. Yapo, 1999. Status of automatic calibration for hydrologic models: Comparison with multilevel expert calibration. *Journal of Hydrologic Engineering* **4**(2): 135–143.

Hargreaves, G.H. and Z.A. Samani, 1985. Reference crop evapotranspiration from temperature. *Applied Engineering in Agriculture* **1**: 96 – 99.

Hill, A.J., and V.S. Neary, 2005. Factors affecting estimates of average watershed slope. *Journal of Hydrologic Engineering* **10**(2): 133–140.

Homer, C., C. Huang, L. Yang, B. Wylie, and M. Coan. 2004. Development of a 2001 national land-cover database for the United States. Ohotogrammetric Engineering & Remote Sensing **70**(7): 829 – 840.

Houston Engineering, 2001. *Summary Report for Activities No. 1 through 7: Wild Rice Watershed District Systems Approach - Finding a Comprehensive Solution to Flooding.* Fargo, N.D., and Maple Grove, Minn.: Houston Engineering, Inc.

Johnson, R.A., and D.W. Wichern. 1998. *Applied Multivariate Statistical Analysis (4th ed.).* Prentice-Hall Inc., New Jersey.

Juracek, K.E., and A.C. Ziegler. 2007. *Estimation of Sediment Sources using Selected Chemical Tracers in the Perry Lake and Lake Wabaunsee basins, Northeast Kansas.* Washington D.C.: the U.S. Department of the Interior and U.S. Geological Survey, Scientific Investigations Report 2007–5020.

Mace, R.E., R.C. Smyth, L. Xu, and J. Liang. 1999. *Transmissivity, Hydraulic Conductivity, and Storativity of the Carrizo-Wilcox Aquifer in Texas.* Austin, Texas: A report prepared for the Texas Water Development Board.

Martinez-Cob, A. and M. Tejero-Juste, 2004. A wind-based qualitative calibration of the Hargreaves ET0 estimation equation in semiarid regions. *Agricultural Water Management* **64**: 251 – 264.

Monteith, J.L., 1965. Evaporation and the environment. p. 205 – 234. *In*: *The State and Movement of Water in Living Organisms, XIXth Symposium.* Society for Experimental Biology, Swansea, Cambridge University Press.

Moore Engineering Inc., 1982. *Engineering Report: Goose River Watershed Hydrologic Analysis and Flood Retention Study.* A Report for Grand Forks, Nelson, North Cass, Steele and Traill County Water Resources Board, and the Red River Joint Water Resource Board and the North Dakota State Water Commission. West Fargo, North Dakota: Moore Engineering Inc.

Motovilov, Y.G., L. Gottschalk, K. England, and A. Rodhe, 1999. Validation of a distributed hydrological model against spatial observations. *Agriculture and Forest Meteorology* **98** – 99: 257 – 277.

Muleta, M.K., J.W. Nicklow, and E.G. Bekele. 2007. Sensitivity of a distributed watershed simulation model to spatial scale. *Journal of Hydrologic Engineering* **12**(2):163–172.

Narasimhan, B., P.M. Allen, R. Srinivasan, S.T. Bednarz, J. Arnold, and J.A. Dunbar. 2007. Streambank erosion and best management practice simulation using SWAT. *In Proc. of ASABE Fourth Conference on Watershed Management to meet Water Quality Standards and TMDLS (Total Maximum Daily Loads).* San Antonio, Texas, March 10 – 14, 2007.

Nash, J. E., and J. V. Sutcliffe. 1970. River flow forecasting through conceptual models: Part I. A discussion of principles. *Journal of Hydrology* **10**(3): 282-290.

Neitsch, S.L., J.G. Arnold, J.R. Kiniry, J.R. Srinivasan, and J.R. Williams, 2002a. *Soil and Water Assessment Tool – User's Manual (version 2000).* Temple, Texas: Grassland, Soil and Water Research Laboratory, Agricultural Research Service; Blackland Research Center, Texas Agricultural Experiment Station.

Neitsch, S.L., J.G. Arnold, J.R. Kiniry, J.R. Williams, and K.W. King, 2002b. *Soil and Water Assessment Tool – Theoretical Documentation (version 2000).* Temple, Texas: Grassland, Soil and Water Research Laboratory, Agricultural Research Service; Blackland Research Center, Texas Agricultural Experiment Station.

Nordt, L. 2004. Late quaternary alluvial stratigraphy of a low-order tributary in central Texas, USA and its response to climate and sediment supply. *Quaternary Research* **62**: 289 – 300.

Penman, H.L., 1956. Evaporation: An introductory survey. *Netherlands Journal of Agricultural Science* **4**: 7 – 29.

Peschel, J.M., P.K. Haan, and R.E. Lacey. 2006. Influences of soil dataset resolution on hydrologic modeling. *Journal of the American Water Resources Association* **42**(5):1371–1389.

Priestley, C.H.B. and R.J. Taylor, 1972. On the assessment of surface heat flux and evaporation using large-scale parameters. *Monthly Weather Review* **100**: 81 – 92.

Srinivasan, M.S., P. Gérard-Marchant, T.L. Veith, W.J. Gburek, and T.S. Steenhuis. 2005. Watershed scale modeling of critical source areas of runoff generation and phosphorus transport. *Journal of the American Water Resources Association* **41**(2): 361 – 375.

Srinivasan, R., T. S. Ramanarayanan, J. G. Arnold, and S. T. Bednarz, 1998. Large-area hydrologic modeling and assessment: Part II. Model application. *Journal of American Water Resources Association* **34**(1): 91 – 101.

Stoner, J.D., D.L. Lorenz, G. J. Wiche, and R.M. Goldstein, 1993. Red River of the North Basin, Minnesota, North Dakota, and South Dakota. *Water Resources Bulletin* **29**(4): 575-615.

TCEQ. 2007. *Updated Evaluation for the North-Central Texas – Trinity and Woodbine Aquifers – Priority Groundwater Management Study Area*. Austin, Texas: Texas Commission on Environmental Quality.

USDA-NRCS, 1995. *Soil Survey Geographic (SSURGO) Data Base: Data Use Information*. Fort Worth, Texas: National Cartography and GIS Center.

USDA–NRCS. 2004. *National Engineering Handbook*. Washington D.C.: U.S. Department of Agriculture (USDA) Natural Resources Conservation Service (NRCS).

USDA–NRCS. 2005. *Soil Survey Geographic (SSURGO) Database*. *Available at* http://www.ncgc.nrcs.usda.gov/products/datasets/ssurgo/index.html, *accessed on* March 16, 2008.

USDA-SCS, 1993. *State Soil Geographic Data Base (STATSGO)*. Miscellaneous Publication No. 1492. Washington, D.C.: U.S. Government Printing Office.

USEPA, 2003. *Land Use and Land Cover (LULC)*. *Available at* http://edc.usgs.gov/products/landcover/lulc.html. *Accessed on* February 16, 2004.

USGS. 1998. *Extreme Precipitation Depths for Texas, Excluding the Trans-Pecos Region*. Washington D.C.: U.S. Geological Survey Water-Resources Investigations Report 98 – 4099.

USGS. 2001a. National elevation dataset. *Available at* http://gisdata.usgs.net/NED/default.asp. *Accessed on* March 20, 2008.

USGS. 2001b. National hydrography dataset. *Available at* http://nhd.usgs.gov. *Accessed on* March 20, 2008.

USGS. 2005. Ground Water Atlas of the United States: Oklahoma, Texas. Washington, D.C.: U.S. geological Survey Open-File Report HA 730-E.

Van Liew, M.W., T.L. Veith, D.D. Bosch, and J.G. Arnold. 2007. Suitability of SWAT for the conservation effects assessment project: A comparison on USDA-ARS experimental watersheds. *Journal of Hydrologic Engineering* **12**(2): 173 – 189.

Vazquez-Amábile, G.G. and B.A. Engel, 2005. Use of SWAT to compute groundwater table depth and streamflow in the Muscatatuck River watershed. *Transactions of the ASAE* **48**(3): 991 – 1003.

Wang, X., and A.M. Melesse. 2005. Evaluation of the SWAT models snowmelt hydrology in a Northwestern Minnesota watershed. *Transactions of the ASAE* **48**(4):1359–1376.

Wang, X., and A.M. Melesse. 2006. Effects of STATSGO and SSURGO as inputs on SWAT model's snowmelt simulation. *Journal of the American Water Resources Association* **42**(5):1217–1236.

Western, A.W., R.B. Grayson, and G. Blöschl. 2002. Scaling of soil moisture: A hydrologic perspective. *Annual Review of Earth and Planetary Sciences* **30**:149–180.

Wiche, G.J., and T. Williams-Sether. 1997. *Water Resources of North Dakota: Streamflow Characteristics of Streams in the Upper Red River of the North Basin, North Dakota, Minnesota, and South Dakota.* Washington, D.C.: U.S. Geological Survey Open-File Report 97 – 416.Williams, J.R. 1980. SPNM, a model for predicting sediment, phosphorus, and nitrogen yields from agricultural basins. *Water Resources Bulletin* **16**: 843 – 848.

Williams, J.R. 1980. SPNM, a model for predicting sediment, phosphorus, and nitrogen yields from agricultural basins. *Water Resources Bulletin* **16**: 843 – 848.

Williams, J.R. 1995. The EPIC model. In *Computer Models of Watershed Hydrology*, ed. V.P. Singh, Ch. 25: 909 – 1000. Highlands Ranch, Colorado: Water Resources Publications.

Windhorn, R.H. 2001. *RAP-M: Rapid Assessment Point Method.* Miscellaneous Pages. Champaign, Illinois: U.S. Department of Agriculture (USDA) Natural Resources Conservation Service (NRCS).

Wright, J.S. 2001. "Desert" loess versus "glacial" loess: Quartz silt formation, source areas and sediment pathways in the formation of loess deposits. *Geomorphology* **36**: 231 – 256.

In: Environmental Modelling: New Research
Editor: Paul N. Findley, pp. 187-204
ISBN: 978-1-60692-034-3

Chapter 8

LANDSLIDE MODELING

Kang-tsung Chang
Kainan University, Taoyuan, Taiwan 33857

Abstract

Because landslide generates a larger yearly loss of property than earthquake, flood, or windstorm, it is important to develop models that can measure the potential of landslide occurrence in an area. Landslide hazard models can be generally grouped into physically-based and statistical models. A physically-based model delineates areas prone to landsliding by analyzing the influence of surface topography on near-surface hydrologic response. It assumes that slope failures are caused by shallow subsurface flow convergence, increased soil saturation, and shear strength reduction. A statistical model predicts the likelihood of landslide occurrence by analyzing the relationship between past landslides and instability factors such as lithology, slope, curvature, aspect, elevation, land use, and drainage. Common statistical methods for landslide prediction include discriminant analysis and logistic regression. Model validation is the process of evaluating a landslide hazard model by comparing model predictions with observed landslides. Observed landslides are compiled from aerial photographs, satellite images, or from ground surveys. This review paper covers landslide mapping, physically-based models, statistical models, model validation, and examples of models for a mountainous watershed in Taiwan.

1. Introduction

The term 'landslide' refers to a downslope movement of a mass of soil and rock material (Cruden 1991; Dikau et al. 1996). Various landslide classifications by morphology, material, mechanism of initiation, and other criteria are available. In general, landslide types include rock falls, rock slides, earth flows, earth slides, and debris flows. Landslides are usually triggered by earthquakes, storms, snow melt, human activities such as road construction, or a combination of these factors (Aleotti and Chowdhury 1999). It has been estimated that landslide generates a yearly loss of property larger than that from any other natural disaster including earthquake, flood, and windstorm (Guzzetti et al. 1999). This is why landslide modeling has generated a great deal of interests in the research community.

A landslide hazard model measures the potential of landslide occurrence within a given area (Varnes 1984; O'Hare and Rivas 2005). The model may be based on explicit probabilities of occurrence or some ordinal categories of probability (e.g., low, moderate, high). Maps of the latter case are usually called 'landslide susceptibility' models (Hansen 1984).

For the past two decades, developments of landslide hazard models have taken advantage of geographic information systems (GIS) (Dikau et al. 1996; Guzzetti et al. 1999; Dai and Lee 2003; Chau et al. 2004; Wang et al. 2005). A GIS is a computer system for capturing, storing, querying, analyzing, and displaying geospatial data (Chang 2007). A GIS is therefore useful for preparing landslide data and layers of instability factors (i.e., factors that can influence landslide occurrence), performing data analysis such as overlay and buffering, and displaying results. Given the important role of GIS, some researchers have added 'GIS-based' to their landslide hazard models (Ayalew et al. 2004; Ayalew and Yamagishi 2005; Lan et al. 2004; Lee 2005; Lee and Choi 2004; Lee et al. 2004; Saha et al. 2005).

After a brief discussion of landslide delineation and mapping, this paper presents two types of landslide hazard models, physically-based and statistical models. It discusses in detail the background and parameters of these models. It also discusses model validation by comparing model predictions with observed landslides. The paper then describes examples of landslide hazard models from Taiwan, an island with frequent landslides triggered by typhoons (tropical hurricanes) and earthquakes. The paper concludes with a short summary of the advantages and disadvantages of physically-based and statistical landslide models.

2. Landslide Mapping

The first step in developing a landslide hazard model, either physically-based or statistical, is to delineate and map past and active landslides in the study area. The locations of these landslides can provide data for model building and for model validation. Traditionally, aerial photographs are used to identify and delineate landslides and to produce landslide inventory maps. Morphologies typical of slope movements such as scarps, deposit zones, disturbed vegetation, and disturbed channels or roads provide visual cues that are important for manually interpreting landslides on an aerial photograph (Dikau 1999). For example, Chang et al. (2008) mapped landslides triggered by typhoon events in Taiwan by comparing ortho-rectified aerial photographs taken before and after each event. Likewise, Galli et al. (2008) compiled three landslide inventory maps for the past 20 years in the Umbria region of Italy by interpreting vertical aerial photographs.

Aerial photographs are not the only source for landslide mapping. Ground survey is often necessary for mapping small landslides under a forest canopy (e.g., Montgomery et al. 2000). In recent years, satellite images, especially those of high-resolution, have also been used for landslide mapping. Nichols and Wong (2004) tested landslide change detection using SPOT imagery and reported detection rates up to 70%. However, other studies have not shown the same degree of success. Using Landsat 7 ETM+ imagery for landslide classification, Petley et al. (2002) found that the classification missed more than 75% of landslides checked by ground surveys in the upland areas of Nepal and Bhutan. Dadson et al. (2004) reported a significant underestimation of small landslides by using 20-m SPOT-4 imagery for landslide mapping. Borghuis et al. (2007) tested automated classification methods of SPOT-5 products for mapping typhoon-triggered landslides and found many errors of commission (i.e., misinterpretation of stable areas as landslides) attributable to the presence of roads, riverbeds, and bare farm fields in the scene.

Regardless of how they are delineated and mapped, landslides are usually transferred onto a grid to develop a model. Each cell on the grid is either a landslide cell or a stable-area cell. An alternative to the grid cell representation is terrain units, area units with homogeneous terrain characteristics. For example, slope units derived from a special software package have been used by Carrara et al. (1991) and others (e.g., Guzzetti et al. 2006) for landslide modeling.

3. Physically-Based Landslide Models

Previous studies on physically-based, or process-based, models have concentrated on deriving the factor of safety (FS) (Dietrich et al. 1995; Pack et al. 1998) and minimum steady-state rainfall (Montgomery and Dietrich 1994; Montgomery et al. 1998; Borga et al. 1998, 2002; Claessens et al. 2007a, b). The FS is an index that determines the slope stability of a grid cell or a terrain unit. And the minimum steady-state threshold is a rainfall rate (mmday^{-1}) critical to induce slope failures. Closely related, the two approaches identify potential landslide areas by using topographic and soil parameters, along with rainfall data. Ideally, predicted landslides should reflect local topographic and soil characteristics as well as local rainfall conditions such as rainfall intensity and duration (Crosta 1998).

3.1. Factor of Safety (FS)

Based on the infinite slope model, the FS is defined as the ratio of the available shear strength (stabilizing forces), including soil and root cohesion, to the shear stress (destabilizing forces) (Pack et al. 1998):

$$FS = \frac{C_r + C_s + \cos^2\theta[\rho_s g(D - D_w) + (\rho_s g - \rho_w g)D_w]\tan\phi}{D\rho_s g\sin\theta\cos\theta} \qquad (1)$$

where C_r [Nm^2] is root cohesion, C_s [Nm^2] is soil cohesion, ρ_s [kgm^{-3}] is wet soil density, ρ_w [1.0 kgm^{-3}] is the density of water, g [9.81 ms^{-2}] is gravitational acceleration, D [m] is the vertical soil depth, D_w [m] is the vertical height of the water table within the soil layer, θ [deg] is slope angle, and ϕ [deg] is the internal friction angle of the soil. In the equation, increasing soil and root strength will increase FS, and increasing slope and groundwater-soil depth ratio (e.g. wetness) will decrease the FS. In principle, FS >= 1 indicates a slope in stable/equilibrium and FS < 1 an unstable slope.

In a pioneer work, Montgomery and Dietrich (1994) combined the infinite slope model with a steady-state hydrologic model to delineate areas prone to landsliding. Their model emphasizes the influence of surface topography on near-surface hydrologic response through shallow subsurface flow convergence, increased soil saturation, shear strength reduction, and landslide generation.

With the inclusion of the hydrologic model, Eq. (1) can be modified in several steps. First, the hydrologic model in Montgomery and Dietrich (1994) interprets the soil thickness (h [m]) and the saturated zone thickness (h_w [m]) as specified perpendicular to the slope rather than measured vertically. Therefore,

$$h = D\cos\theta \tag{2}$$

$$h_w = D_w\cos\theta$$

Second, the relative wetness, w [-], is the ratio of local flux at a given steady state rainfall to that at soil profile saturation:

$$w = h_w/h \tag{3}$$

Further, based on the work of O'Laughlin (1986) in predicting surface saturation zones, w can be defined as:

$$w = Ra/bT\sin\theta \tag{4}$$

where R [md^{-1}] is the net rainfall (precipitation less evapotranspiration and deep drainage losses), a [m^2] is the upslope contributing drainage area, T [m^{2}d^{-1}] is the soil transmissivity. The unit contour length, b [m], is added to the equation to make w dimensionless. And according to Pack et al. (2001), the grid resolution can be taken as the effective contour length. The use of Eq. (4) also assumes that (1) the direction of subsurface flow is locally parallel to the gradient of surface topography, and (2) the saturation zone is in equilibrium with the steady state rainfall. The second assumption implies that w cannot exceed 1 because any excess forms overland flow. Therefore w ranges between 0 and 1.

Third, another dimensionless term C [-] can also be used. C is combined cohesion, defined as the ratio of total cohesive strength (root cohesion plus soil cohesion) relative to the soil depth (Pack et al. 1998):

$$C = (Cr + Cs) / (h\,\rho_s\,g) \tag{5}$$

Substituting Eqs (3)-(5) in Eq. (1), the FS can be expressed as:

$$FS = \frac{C + \cos\theta\left[1 - \min\left(\frac{Ra}{bT\sin\theta}, 1\right)\left(\frac{\rho_w}{\rho_s}\right)\right]\tan\phi}{\sin\theta} \tag{6}$$

3.2. Critical Rainfall Model

By equating the FS in Eq. (6) to 1, which is the threshold for instability, and solving for *R*, we can determine the critical rainfall Q_{cr} [mday^{-1}], predicted to cause slope failure:

$$Q_{cr} = T \sin\theta \left(\frac{b}{a}\right)\left(\frac{\rho_s}{\rho_w}\right)\left[1 - \frac{(\sin\theta - C)}{(\cos\theta \tan\phi)}\right] \quad (7)$$

Once Q_{cr} is crossed in a unit area, the unit remains unstable at greater rainfall rates.

Given the boundary conditions for *w* (between 0 and 1), the conditions for upper and lower thresholds for possible slope failures can also be defined. Unconditionally stable areas are areas predicted to be stable even when saturated and satisfy:

$$\tan\theta \le \left(\frac{C}{\cos\theta}\right) + \left(1 - \frac{\rho_w}{\rho_s}\right)\tan\phi \quad (8)$$

Unconditionally unstable areas are areas predicted to be unstable even when dry and satisfy:

$$\tan\theta > \tan\phi + \left(\frac{C}{\cos\theta}\right) \quad (9)$$

4. Statistical Landslide Model

A common approach to assessing the probabilities of occurrence of landslides is to use information of past landslides (Carrara et al. 1991; Chau et al. 2004). Given a landslide inventory map, a variety of statistical methods have been proposed for modeling landslides (Aleotti and Chowdhury 1999; Guzzetti et al. 1999; Ercanoglu and Gokceoglu 2004; Wang et al. 2005). Statistical methods treat the landslide mechanism as a 'black box' and assess landslide hazard by analyzing the functional relationships between the instability factors and the distribution of recorded landslides. Statistical methods can be bivariate or multivariate.

4.1. Bivariate Analysis

Bivariate analysis uses favorability functions (Chung and Fabbri 1993) to create a landslide susceptibility map from a selection of instability factors. By comparing a landslide inventory map with the map of an instability factor, the analysis first transforms different classes of the factor into a numeric scale representing the likelihood or certainty of landslide occurrence. Thus, a slope map can be transformed into a map of landslide susceptibility values for different slope classes. Then favorability values for each instability factor are integrated according to a specific rule or integration function to create a landslide susceptibility model.

The certainty factor (CF) is one type of bivariate analysis (Luzi and Pergalani 1999; Remondo et al. 2003). The CF is defined as a function of probability:

$$\mathrm{CF} = \frac{pp_a - pp_s}{pp_a(1 - pp_s)}, \text{ if } pp_a >= pp_s \qquad (10)$$
$$= \frac{pp_a - pp_s}{pp_s(1 - pp_a)}, \text{ if } pp_a < pp_s$$

where pp_a is the conditional probability of having a number of landslide events occurring in class a and pp_s is the prior probability of having the total number of landslide events occurring in the study area. For example, to calculate pp_a for the slope class of 40-50^0, we will derive the area of landslides falling within the slope class and divide the area by the total area of the slope class. And, to calculate pp_s, we will divide the total area of landslides by the size of the study area. The range of the CF is [-1, 1]: a positive value means an increasing certainty in landslide occurrence, while a negative value a decreasing certainty. A value close to 0 means the prior probability is very similar to the conditional one so that it is difficult to give any indication about the certainty of landslide occurrence.

After the CF values of each class for all layers are calculated, the layers are combined pairwise. The combination of two CFs, x and y, is expressed as z as follows:

$$z = \begin{cases} x + y\,xy,\ if\, x, y >= 0 \\ \dfrac{x + y}{1 - \min(|x|, |y|)},\ if\, x, y \text{ opposite sign} \\ x + y + xy,\ if\, x, y < 0 \end{cases} \qquad (11)$$

The result from combining or integrating all layers is a landslide hazard model.

4.2. Multivariate Analysis

A variety of multivariate methods for landslide modeling exist. Two common methods are discriminant analysis and logistic regression. Discriminant analysis uses a linear function and a set of variables (instability factors) to classify samples into alternative groups (landslide or stable area):

$$L = B_0 + B_1X_1 + B_2X_2 + B_3X_3 + \ldots + B_nX_n \qquad (12)$$

where L is the presence or absence (or probability) of landslide in a cell or unit area, the X's are the input variables, and the B's are the coefficients. Discriminant analysis has been used for landslide modeling by Carrara et al. (1991, 2003), Guzzetti et al. (2006), and Galli et al. (2008).

In recent years, many researchers have used logistic regression to predict probabilities of landslide occurrence by analyzing the functional relationships between the instability factors and the past distribution of landslides (Guzzetti et al. 1999; Dai and Lee 2003; Ohlmacher and Davis 2003; Ayalew and Yamagishi 2005; Can et al. 2005; Wang et al. 2005; Yesilnacar and Topal 2005). The assumption is that factors, which caused landslides in the past, are the same as those, which will trigger landslides in the future.

Logistic regression is useful when the dependent variable is categorical (e.g., presence or absence) and the explanatory (independent) variables are categorical, numeric, or both (Menard 2002). Compared to discriminant analysis, logistic regression has the advantage of being less sensitive to deviances from normality of the input variables (Carrara et al. 2008). The logit model has the following form:

$$\text{logit}\ (y) = a + b_1x_1 + b_2x_2 + b_3x_3 + \ldots + e \tag{13}$$

where the logit of y is the dependent variable, x_i is the explanatory variable i, a is a constant, b_i is the regression coefficient i, and e is the error term. The logit of y is the natural logarithm of the odds:

$$\text{logit}\ (y) = \ln(\, p/(1 - p)) \tag{14}$$

where p is the probability of the occurrence of y and $p/(1 - p)$ is the odds. To convert logit (y) back to the probability p, Eq. (14) can be rewritten as:

$$p = \frac{\exp(a + b_1x_1 + b_2x_2 + b_3x_3 + \ldots)}{1 + \exp(a + b_1x_1 + b_2x_2 + b_3x_3 + \ldots)} \tag{15}$$

A logit model can be evaluated by the receiver operating characteristic (ROC) curve, which is based on the proportions of true positive (proportion of incidences correctly reported as positive) and false positive (proportion of incidences erroneously reported as positive). Typically, a probability value of 0.5 is used to determine whether the model has made a correct prediction (> 0.5) or not (< 0.5). The area under the ROC curve (AUC) can measure the fitness of a model: the larger the area, the better the model. Additionally, Cox and Snell R^2 and Nagelkerke R^2 measure how well the explanatory variables can predict and explain the dependent variable. Cox and Snell R^2 cannot achieve a maximum of 1, whereas Nagelkerke R^2 stretches the R^2 value to range from 0 to 1.

For a logit model of landslides, the dependent variable separates landslide (1) from stable area (0) and the explanatory variables typically include geologic and geomorphic factors. For example, Dai and Lee (2002) used lithology and geological structure, slope gradient and morphology, aspect, elevation, land-use type, and proximity to drainage line as explanatory variables for their study of landslide characteristics in Hong Kong. Although many shallow landslides are triggered by rainfall, few studies have used rainfall as an explanatory variable primarily because of the lack of rainfall data, especially in mountainous watersheds. Dai and Lee (2003) included rolling 24-hr rainfall or 24-hr rainfall isohyets as an explanatory variable in their modeling of storm-induced shallow landsliding. Chang et al. (2008) used maximum 3-hr rainfall intensity and total rainfall duration derived from radar data associated with a typhoon event to develop their landslide logit model.

5. Model Validation

Model validation refers to the process of evaluating the performance of a model by comparing model predictions with observed landslide locations. The observed landslides for model validation (validation set) should not be the same as those used to develop the model (training set). Some researchers insist that the validation set should be separated from the training set temporally or spatially (e.g., Chung and Fabbri 2003). Others recommend that landslides from an inventory map can be split into a training set and a validation set (e.g., Carrara et al. 2008).

Model validation typically starts by overlaying observed landslides with model predictions so that the two can be directly compared. For a robust CF model, a disproportionate percentage of observed landslides would occur in sites predicted to be unstable. Likewise, for a reliable critical rainfall model, a majority of observed landslides would fall within areas classified as unconditionally unstable or low critical rainfall values.

The performance of a discriminant or logistic model can be evaluated by different measures. A simple measure is the 'area concordance' defined as (Borghuis et al. 2007):

[(overlapped area between model predictions and observed landslides) / (total observed landslide area)] * 100 (16)

The area concordance only considers landslide areas. It is an appropriate (and rigorous) measure to use if landslide areas cover only a small portion of the study area. The area concordance concept can be easily extended to include both landslide and stable areas by simply tabulating the overall percent of cells correctly classified as either landslide or stable area (e.g., Carrara et al. 2008). Another measure for including both landslide and stable areas is the Kappa statistic (e.g., Guzzetti et al. 2006; Thiery et al. 2007):

$$k = [P_r(a) - P_r(e)] / [1 - P_r(e)] \quad (17)$$

where $P_r(a)$ is the observed agreement, and $P_r(e)$ is the probability that agreement is due to chance. The Kappa statistic can range from +1 (perfect agreement) via 0 (no agreement above that expected by chance) to -1 (complete disagreement). Suppose a 10-by-10 grid has 4 observed landslide cells, 2 of which are correctly predicted, and 96 observed stable cells, all of which are correctly predicted (Figure 1). The Kappa statistic k for this example can be computed as follows:

$$\text{Observed agreement} = (2+96) / 100 = 0.98$$

$$\text{Chance agreement} = (0.04 \times 0.02) + (0.96 \times 0.98) = 0.9416$$

$$k = (0.98 - 0.9416) / (1 - 0.9416) = 0.658$$

	Landslide	Stable Area	
Landslide	2	0	2 (0.02)
Stable Area	2	96	98 (0.98)
	4 (0.04)	96 (0.96)	

Figure 1. An example for calculating the Kappa statistic.

6. Examples of Landslide Models

The study area is the 120-km^2 Baichi watershed, a sub-watershed of the Shihmen watershed, in northern Taiwan. Elevations in the watershed range from 830 m in the northwest to 3320 m a.s.l. in the southeast, with generally rugged topography. Nearly 90% of the study area is forested. Areas in the remaining 10% are cultivated fields and built-up areas located in lower elevations of the northern part of the watershed. The climate is influenced by typhoons in summer and the northeast monsoon in winter. The mean monthly temperature is 27.5°C in July and 14.2°C in January, with the mean annual temperature of 21°C. The annual precipitation averages 2370 mm. Because of typhoons, large rainfall events usually happen from May to September, triggering many shallow landslides.

6.1. A Critical Rainfall Model

To derive the critical rainfall model, three soil physical characteristics, saturated bulk density (ρ_s), transmissivity (T), and internal friction angle (ϕ), were measured in each of the three lithological formation units in the watershed. To estimate the spatial variation of combined cohesion (C), the NDVI (normalized difference vegetation index) values were first retrieved from the 8-m FORMOSAT-2 satellite images and then linearly transformed to values ranging from 0.0 to 50.0 kpa (Huang et al. 2006). The local slope angle (θ) and the upslope contributing drainage area (a) were calculated from a 10-m digital elevation model (DEM) compiled from the stereo pairs of 1:5000 aerial photographs.

Figure 2 shows the critical rainfall model including unconditionally unstable area, six critical rainfall categories ranging from 0-50 to > 400 mmd^{-1}, and unconditionally stable area. 64.3% of the total area is unconditionally stable, and 6.0% is unconditionally unstable. To assess the performance of the model, we delineated and mapped landslides triggered by Typhoon Aere (August 23-25, 2004) by comparing ortho-rectified aerial photographs taken before (August 6, 2004) and after (September 2, 2004) the typhoon. Typhoon Aere triggered

421 new landslides. Most observed slope failures were shallow landslides on soil mantled slopes with depth less than 2 m. 51.5% of these landslides occurred in unconditionally unstable cells and 24.7% in cells of the lowest critical rainfall range (0-50 $mmday^{-1}$).

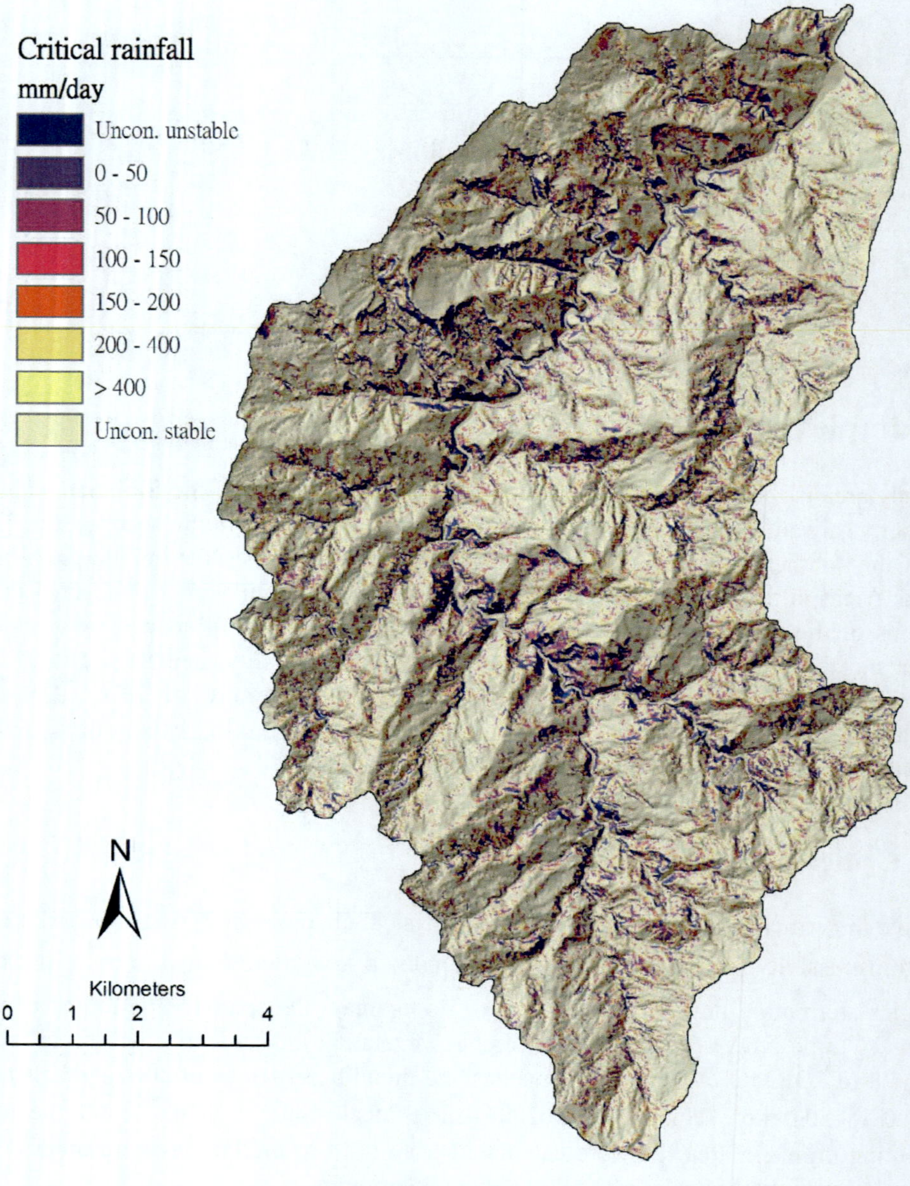

Figure 2. The critical rainfall distribution map.

6.2. A Certainty Factor Model

Data for building and validating the CF model consisted of landslides and instability factors for the Baichi watershed. Landslide inventory maps from Taiwan's Soil and Water

Conservation Bureau record 219 landslides in 1986, 34 in 1992, 468 in 2004, and 517 in 2006. The agency describes landslides in the inventory as active landslides for the recorded year but does not further differentiate them as new, re-activated, or old landslides. For this study, landslides from 1986, 1992, and 2004 were included in the training set, and landslides from 2006 in the validation set. Table 1 shows the descriptive statistics of these landslide areas and the percentage of total landslide area within the study area.

Table 1. Descriptive statistics of landslide areas from 1986, 1998, 2004, and union, and percentage of total landslide area within the study area. (Landslide areas measured in sq m^2)

Year	Count	Minimum	Maximum	Mean	Stan. Dev.	Total Landslide Area	% Study Area
1986	219	2117	235076	14073	19424	3082020	2.55
1992	34	1597	64632	14693	16700	499547	0.41
2004	468	67	459252	6586	23785	3082069	2.55
Training set	716	18	886842	10760	37305	7703904	6.38
2006 (Validation set)	517	18	661143	5279	30002	2729268	2.26

Table 2. CF value for each category of the instability factors.

Slope (in degrees)									
Category	0-10	10-20	20-30	30-40	40-50	50-60	60-90		
CF	0	-0.4835	-0.4487	-0.1751	0.2516	0.5245	0.7719		
Aspect									
Category	Flat	N	NE	E	SE	S	SW	W	NW
CF	0.1475	-0.0789	-0.1859	-0.0441	0.0737	0.3009	0.1880	-0.1697	-0.0588
Lithology									
Category	10	20	30						
CF	0.0689	-0.2488	0.3774						
Elevation (in meters)									
Category	<1383	1383-1658	1658-1886	1886-2096	2096-2351	2351-3375			
CF	0.4957	0.2374	0.0073	-0.2827	-0.5839	-0.3537			
Distance to Stream									
Category	0-50m	50-100m	>100m						
CF	0.6975	0.5682	-0.3872						
Wetness Index									
Category	<0.1	0.1-0.5	0.5-1.0	1.0-3.0	3.0-5.0	>5.0			
CF	-0.1207	-0.0202	-0.0588	0	0.2825	0.4130			
Soil									
Category	1	2	3	4	5				
CF	-0.7558	-0.2540	0.2121	-0.0072	0.6613				
Distance to Fault (in km)									
Category	<1	1-2	2-3	3-4	4-5	>5			
CF	-0.2326	0.3369	0.0441	-0.1552	-0.9343	-0.5942			

The instability factors included slope, aspect, lithology, elevation, distance to stream, wetness index, soil, and distance to fault line. Lithology and faults were derived from 1:10,000 geology maps; streams and roads from 1:25,000 topographic maps; soils from 1:5,000 soil maps; and slope, aspect, and wetness from 9-m DEMs. The northeastern part of the watershed (an aboriginal protection district) was not included in the model because no soil data were available for the area. Table 2 shows the CF values by category for each of the eight instability factors.

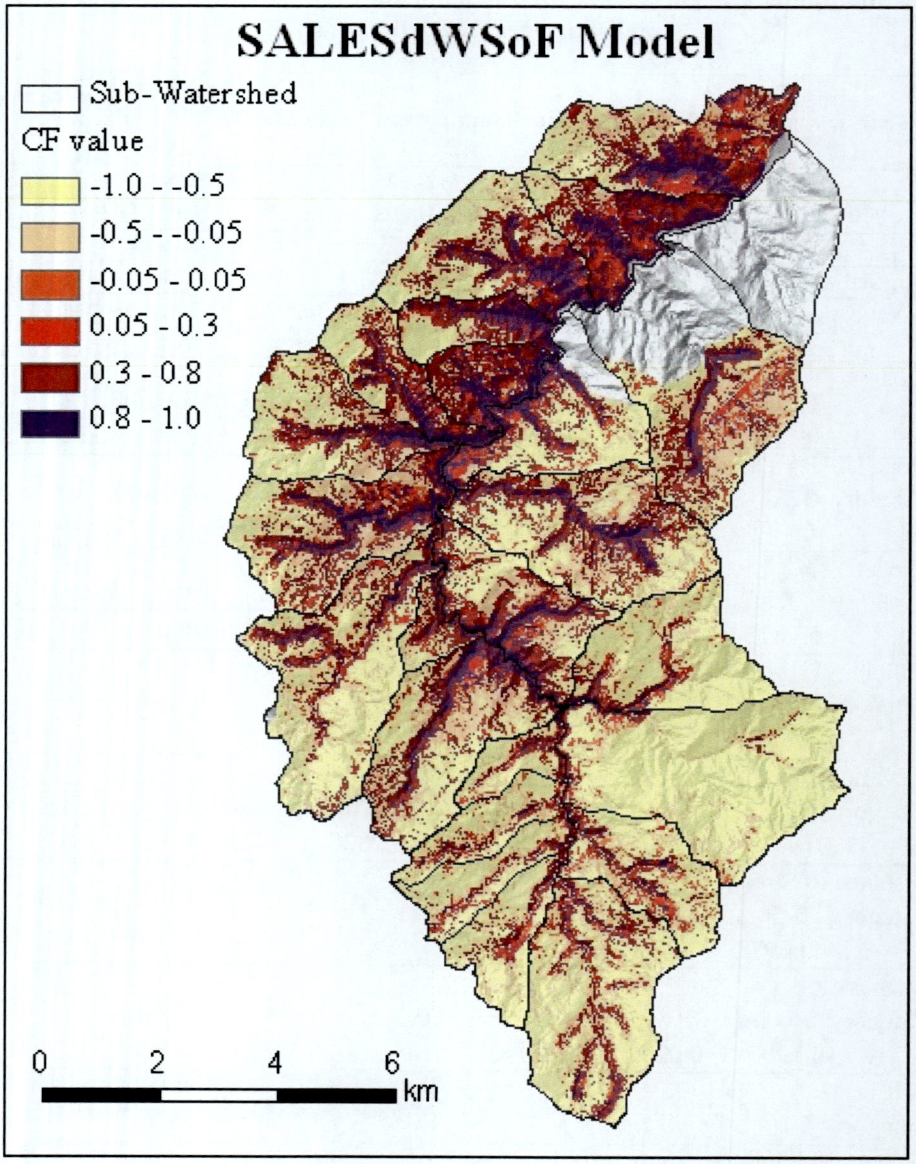

Figure 3. The CF distribution map based on the overlay of slope (S), aspect (A), lithology (L), elevation (E), distance to stream (Sd), wetness index (W), soil (So), and distance to fault (F).

After all instability factors were integrated, the integrated CF values were classified into six hazard classes: high stability (-1.0 - -0.5), medium stability (-0.5 - -0.05), uncertainty (-0.05 – 0.05), low instability (0.05 – 0.3), medium instability (0.3 – 0.8), and high instability (0.8 – 1.0). Figure 3 shows the CF distribution map of the final combined layer.

To assess the performance of the model, we overlaid the 2006 landslide map, with the final combined layer. Then we calculated percent landslide area within each hazard class and percent total area for each hazard class. Figure 4 shows the results. Class 6 (high instability) covers 37% of landslide area but 12% of total area and class 5 (medium instability) covers 25% of landslide area but 18% of total area. A disproportionate percentage of observed landslides did occur in sites predicted to be unstable; however, Class 1 (high stability) and Class 2 (medium stability) together still cover almost 30% of landslide area.

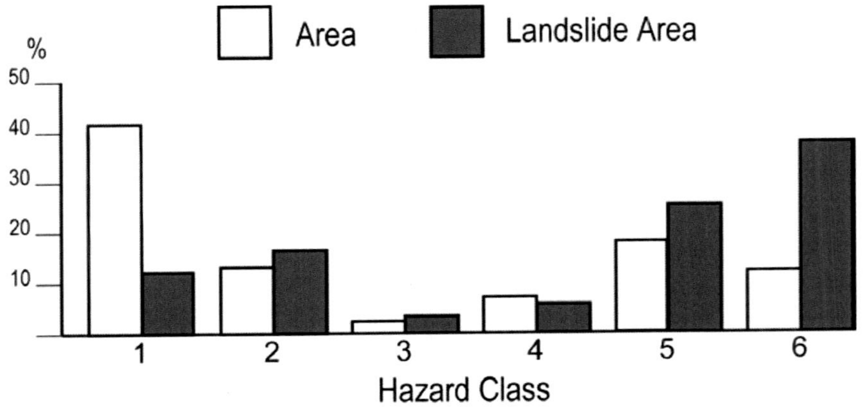

Figure 4. Validation of the CF model by comparing the percentage of area with the percentage of landslide area for each hazard class.

6.3. A Logistic Regression Model

Both the study area and the data (landslides and instability factors) for the logit model were the same as those for the CF model. The model is significant at the 0.01 level, with Cox & Snell $R^2 = 0.25$, Nagelkerke $R^2 = 0.34$, and AUC = 0.72. The explanatory variables that are significant at the 0.01 level include: aspect (south), distance to fault, elevation (1383-1886 m and 2356-3375 m), lithology, slope (> 20°), soil, distance to stream, and wetness index (> 1.0). Figure 5 displays the landslide probability map based on the logit model.

To assess model performance, a cell in the model output was classified as a landslide cell if its probability of landslide occurrence was 0.5 or greater and a stable cell if its probability was less than 0.5. The area concordance between the 2006 landslide map and the model output turned out to be 72%.

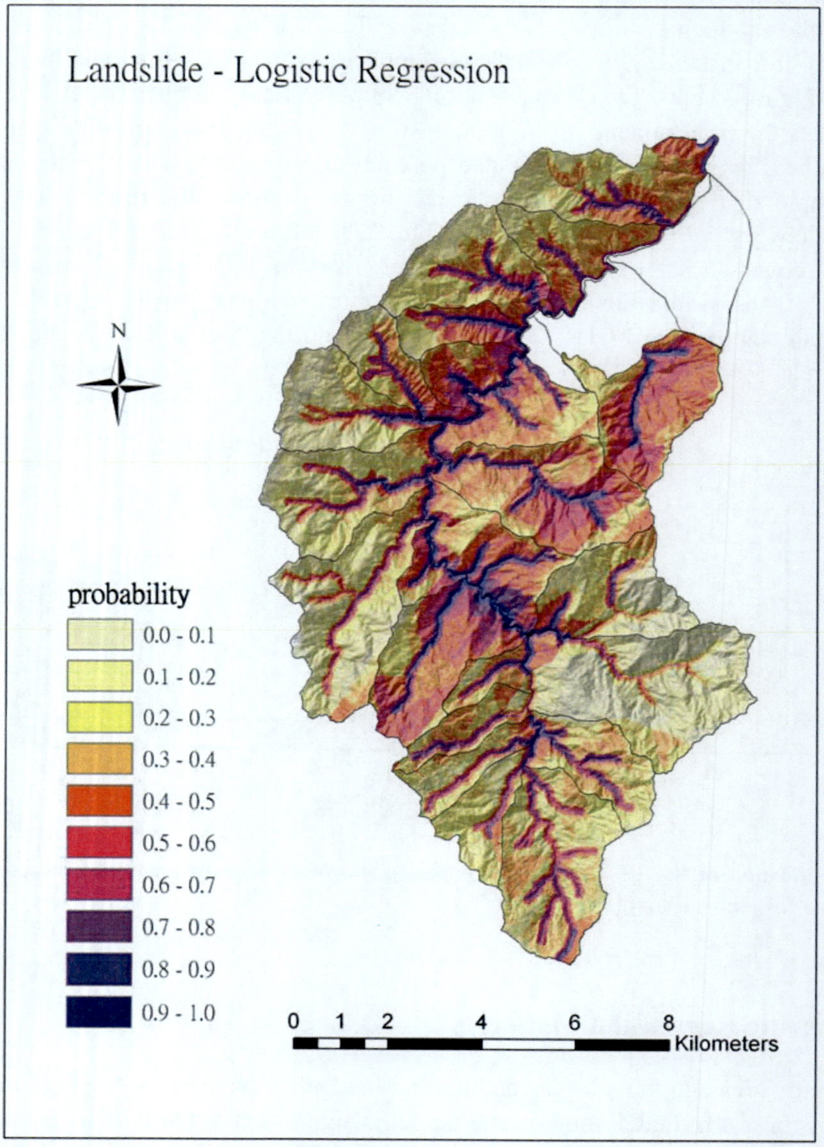

Figure 5. The probability map of the logistic regression model.

7. Conclusion

This paper has discussed physically-based models and statistical models for predicting landslide occurrence. Physically-based models identify potential landslide areas by analyzing the influence of surface topography on near-surface hydrologic response. A physically-based model requires local topographic and soil parameters, in addition to local rainfall data. It can be difficult to gather all these input data, especially the physical and mechanical soil properties, over large and complex areas. In many cases, estimated data or surrogate variables (e.g., NDVI for combined cohesion) are used for model building, causing uncertainty of varying degrees in the model result.

Statistical models treat the landslide mechanism as a 'black box' and assess landslide hazard by assuming that instability factors that have caused slope failures in the past are the same as those that will generate landslides in the future. This assumption is appropriate if the interactions of a large set of instability factors that trigger landslides are unclear or unknown. Because instability factors included in a statistical model are specific to a region, the model cannot be used in other regions, thus limiting its applicability.

As a natural disaster, landslide incurs a larger loss of property yearly than earthquake, flood, and windstorm. Thus, we can safely predict that researchers will continue to employ physically-based models to better understand the mechanism of landsliding and statistical models to assess landslide hazard at the regional level. At the same time, a reliable temporal inventory of landslides is needed for model building and validation.

References

Aleotti, P.; Chowdhury, R. *B. Eng. Geol. Environ.* 1999, 58, 21-44. Landslide hazard assessment: summary review and new perspectives.

Ayalew, L.; Yamagishi, H.; Ugawa, N. *Landslides* 2004, 1, 73-81. Landslide susceptibility mapping using GIS-based weighed linear combination, the case in Tsugawa area of Agano River, Niigata Prefecture, Japan.

Ayalew, L.; Yamagishi, H. *Geomorphology* 2005, 65, 15-31. The application of GIS-based logistic regression for landslide susceptibility mapping in the Kakuda-Yahiko Mountains, Central Japan.

Borga, M.; Dalla Fontana, G.; Da Ros D.; Marchi, L. *Environ. Geol.* 1998, 35, 81-88. Shallow landslide hazard assessment using a physically based model and digital elevation data.

Borga, M.; Dalla Fontana, G.; Gregoretti, C.; Marchi, L. *Hydrol. Process* 2002, 16, 2833-2851. Assessment of shallow landsliding by using a physically based model of hillslope stability.

Borghuis, A.M.; Chang, K.; Lee, H.Y. *Int. J. of Remote Sens.* 2007, 28, 1843-1856. Comparison between automated and manual mapping of typhoon-triggered landslides from SPOT-5 imagery.

Can, T.; Nefeslioglu, H.A.; Gokceoglu, C.; Sonmez, H.; Duman, T.Y. *Geomorphology* 2005, 72, 250-271. Susceptibility assessments of shallow earthflows triggered by heavy rainfall at three catchments by logistic regression analyses.

Carrara, A.; Cardinali, M.; Detti, R.; Guzzetti, F.; Pasqui, V.; Reichenbach, P. *Earth Surf. Proc. Land.* 1991, 16, 427-445. GIS techniques and statistical models in evaluating landslide hazard.

Carrara, A.; Crosta, G.; Frattini, P. *Earth Surf. Proc. Land.* 2003, 28, 1125-1142. Geomorphological and historical data in assessing landslide hazard.

Carrara, A.; Crosta, G.; Frattini, P. *Geomorphology* 2008, 94, 353-378. Comparing models of debris-flow susceptibility in the alpine environment.

Chang, K. *Introduction to Geographic Information Systems*; McGraw-Hill: New York, 2007.

Chang, K.; Chiang, S.; Feng, L. *Earth Surf. Proc. Land.* 2008, 33, 1261-1271. Analysing the relationship between typhoon-triggered landslides and critical rainfall conditions.

Chau, K.T.; Sze, Y.L.; Fung, M.K.; Wong, W.Y.; Fong, E.L.; Chan, L.C.P. *Compu. Geosci.* 2004, 30, 429-443. Landslide hazard analysis for Hong Kong using landslide inventory and GIS.

Chung, C.F.; Fabbri, A.G. *Nonrenew. Resour.* 1993, 2, 122-139. The representation of geoscience information for data integration.

Chung, C.F.; Fabbri, A.G. *Nat. Hazards* 2003, 30, 451-472. Validation of spatial prediction models for landslide hazard mapping.

Claessens, L.; Schoorl, J.M.; Veldkamp, A. *Geomorphology* 2007a, 87, 16-27. Modelling the location of shallow landslides and their effects on landscape dynamics in large watersheds: An application for northern New Zealand.

Claessens, L.; Knapen, A.; Kitutu, M.G.; Peosen, J.; Deckers, J.A. *Geomorphology* 2007b, 90, 23-25. Modelling landslide hazard, soil redistribution and sediment yield of landslides on the Ugandan footslopes of Mount Elgon.

Crosta, G. *Environ. Geol.* 1998, 35, 131-145. Regionalization of rainfall thresholds: an aid to landslide hazard evaluation.

Cruden, D. M. *B. Int. Assoc.Eng. Geol.* 1991, 43, 27-29. A simple definition of a landslide.

Dadson, S.J.; Hovius, N.; Chen, N.; Dade, B.W.; Lin, J.C.; Hsu, M.L.; Lin, C.W.; Horng, M.J.; Chen, T.C.; Milliman, J.; Stark, C.P. *Geology* 2004, 32, 733-736. Earthquake-triggered increase in sediment delivery from an active mountain belt.

Dai, F.C.; Lee, C.F. *Geomorphology* 2002, 42, 213-228. Landslide characteristics and slope instability modeling using GIS, Lantau Island, Hong Kong.

Dai, F.C.; Lee, C.F. *Earth Surf. Proc. Land.* 2003, 28, 527-545. A spatiotemporal probabilistic modeling of storm-induced shallow landsliding using aerial photographs and logistic regression.

Dietrich, W.E.; Reiss, R.; Hsu, M.; Montgomery, D.R. *Hydrol. Process* 1995, 9, 383-400. A process-based model for colluvial soil depth and shallow landsliding using digital elevation data.

Dikau, R.; Brunsden, D.; Schrott, L.; Ibsen, M.; Eds.; *Landslide Recognition: Identification, Movement, and Causes*; Wiley: New York, 1996.

Ercanoglu, M.; Gokceoglu, C. *Eng. Geol.* 2004, 75, 229-250. Use of fuzzy relations to produce landslide susceptibility map of a landslide prone area (West Black Sea Region, Turkey).

Galli, M.; Ardizzone, F.; Cardinali, M.; Guzzetti, F.; Reichenbach, P. *Geomorphology* 2008, 94, 268-289. Comparing landslide inventory maps.

Guzzetti, F.; Carrara, A.; Cardinali, M.; Reichenbach, P. *Geomorphology* 1999, 31, 181-216. Landslide hazard evaluation: a review of current techniques and their application in a multi-scale study, Central Italy.

Guzzetti, F.; Reichenbach, P.; Ardizzone, F.; Cardinali, M.; Galli, M. *Geomorphology* 2006, 81, 166-184. Estimating the quality of landslide susceptibility models.

Hansen, A.; In *Slope Stability*; Brunsden, D.; Prior, D. B.; Ed.; Landslide hazard analysis; Wiley: New York, 1984; pp. 523-602.

Huang, J.; Kao, S.; Hsu, M.; Lin, J. *Nat. Hazards Earth Syst. Sci.* 2006, 6, 803-815. Stochastic procedure to extract and to integrate landslide susceptibility maps: an example of mountainous watershed in Taiwan.

Lan, H.X.; Zhou, C.H.; Wang, L.J.; Zhang, H.Y.; Li, R.H. *Eng. Geol.* 2004, 76, 109-128. Landslide hazard spatial analysis and prediction using GIS in the Xiaojiang watershed, Yunnan, China.

Lee, S. *Int. J. of Remote Sens.* 2005, 26,1477-1491. Application of logistic regression model and its validation for landslide susceptibility mapping using GIS and remote sensing data.

Lee, S.; Choi, J. *Int. J. .Geogr. Inf. Sci.* 2004, 18, 789-814. Landslide susceptibility mapping using GIS and the weight-of-evidence model.

Lee, S.; Choi, J.; Min, K. *Int. J. of Remote Sens.* 2004, 25, 2037-2052. Probabilistic landslide hazard mapping using GIS and remote sensing data at Boun, Korea.

Luzi, L.; Pergalani, F. *Nat. Hazards* 1999, 20, 57-82. Slope instability in static and dynamic conditions for urban planning: the 'Oltre Po Pavese' case history (Regione Lombardia—Italy).

Menard, S., 2002. Applied Logistic Regression Analysis, 2d ed. Sage, Thousand Oaks, CA.

Montgomery D.R.; Dietrich W.E. *Water Resour. Res.* 1994, 30, 1153-1171. A physically based model for topographic control on shallow landsliding.

Montgomery, D.R.; Sullivan, K.; Greenberg, H.M. *Hydrol. Process.* 1998, 12, 943-955. Regional test of a model for shallow landsliding.

Montgomery, D.R.; Schmidt, K.M.; Greenberg, H.M.; Dietrich, W.E. *Geology* 2000, 28, 311-314. Forest clearing and regional landsliding.

Nichol, J.; Wong, M.S. *Int. J. of Remote Sens.* 2005, 26, 1913-1926. Satellite remote sensing for detailed landslide inventories using change detection and image fusion.

O'Hare, G.; Rivas, S. *Geogr. J.* 2005, 171, 239-258. The landslide hazard and human vulnerability in La Paz City, Bolivia.

Ohlmacher, G.C.; Davis, J.C. *Eng. Geol.* 2003, 69, 331-343. Using multiple logistic regression and GIS technology to predict landslide hazard in northeast Kansas, USA.

O'Loughlin, E.M. *Water Resour. Res.* 1986, 22, 794-804. Prediction of surface saturation zones in natural catchments by topographic analysis.

Pack, R.T.; Tarboton, D.G.; Goodwin, C.N. 1998. The SINMAP approach to terrain stability mapping. The 8th Congress of the International Association of Engineering Geology, Vancouver, British Columbia.

Pack, R.T.; Tarboton, D.G.; Goodwin, C.N. 2001. Assessing terrain stability in a GIS using SINMAP. The 15th Annual GIS Conference, GIS 2001, Vancouver, British Columbia.

Petley, D.N.; Crick, W.D.O.; Hart, A.B. (2002). The use of satellite imagery in landslide studies in high mountain area. http://www.gisdevelopment.net/aars/acrs/2002/hdm/48.pdf.

Remondo, J.; Gonzalez-Diez, A.; Diaz de Teran J.; Cendrero, A. *Nat. Hazards* 2003, 30, 267-279. Landslide susceptibility models utilising spatial data analysis techniques. A case study from the lower Deba Valley, Guipuzcoa (Spain).

Saha, A.K.; Gupta, R.P.; Sarkar, I.; Arora, M.K.; Csaplovics, E. *Landslides* 2005, 2, 61-69. An approach for GIS-based statistical landslide susceptibility zonation—with a case study in the Himalayas.

Thiery, Y.; Malet, J.; Sterlacchini, S.; Puissant, A.; Maquaire, O. *Geomorphology* 2007, 92, 38-59. Landslide susceptibility assessment by bivariate methods at large scales: Application to a complex mountainous environment.

Varnes, D.J. *Landslide Hazard Zonation: A Review of Principles and Practice*; UNESCO Press: Paris, 1984.

Wang, H.; Liu, G.; Xu, W.; Wang, G. *Prog. .Phys. Geog.* 2005, 29, 548-567. GIS-based landslide hazard assessment: an overview.

Yesilnacar, E.; Topal, T. *Eng. Geol.* 2005, 79, 251-266.

Landslide susceptibility mapping: a comparison of logistic regression and neural networks methods in a medium scale study, Hendek region (Turkey).

In: Environmental Modelling: New Research
Editor: Paul N. Findley, pp. 205-221
ISBN: 978-1-60692-034-3

Chapter 9

SPATIAL MODELLING OF GROUNDWATER POLLUTION USING A GIS

M. Maanan[1] and M. Robin[2]
Université de Nantes, UMR 6554 LETG–Géolittomer,
B.P 81227 – 44312 Nantes Cedex 3 (France)

Abstract

This study investigates the groundwater contamination by heavy metals of the industrial area of Jorf Lasfar, located on the Atlantic coast of El Jadida, Morocco. This paper describes groundwater mapping and data collection and management to be created. Water samples collected from wells near the area of the industrial installations immediately located on the coast for chemical analyses. Multivariate statistics and GIS techniques were applied to classify the elements, to facilitate interpretation of the spatial relationships among key environmental processes and to identify elements influenced by human activities.

The results show that the area in general is characterized by hard water and high salinity hazard, possibly due to its proximity and hydraulic connection with the sea. Fe, Mn, F, Cu, Pb, Ni, Cr and Cd were found to be the major contaminants in groundwater. The analysis of groundwater indicates contamination at various degrees. Spatial distribution modelling of element concentrations is produced to indicate contamination plumes from possible anthropogenic sources. It was observed that the groundwater in south-eastern of the Jorf Lasfar industrial area is contaminated due to industrial effluents in coast and predominant wind direction. Cluster analysis (CA) classified the elements into two groups: the first group being influenced by human activities, the second predominantly derived from natural sources.

Key words: Groundwater; spatial modelling; GIS; heavy metal; human activities; pollution.

---

[1] E-mail address: mohamed.maanan@univ-nantes.fr
[2] E-mail address: marc.robin@univ-nantes.fr

Introduction

Pollution of groundwater is receiving increased attention from Moroccan regulatory agencies and from water users. As a result, pollution has been found to be much more widespread than we had believed only a few years ago. This attention has also resulted in widespread recognition of the facts that polluted ground water may pose a serious threat to health that is often not apparent to those affected and that purification of polluted ground-water systems may require centuries or the expenditure of huge sums of money. These facts alone make it imperative that the pollution of ground water by harmful substances absolutely be avoided to the maximum possible extent.

Available water resources in Morocco are limited due to the country's geographical location. However, the situation is intensified by population growth and resource wastage. Water pollution and deforestation impose increasing pressure on water quality and undermine ecological equilibrium. In El Jadida coastal areas "*Sahel*", fresh groundwater derived from precipitation on the land comes in contact with and discharges into the sea containing brackish water. However, deterioration of water due to pollution, caused by industries in particular, has a profound impact on the quality of groundwater. Before 1986, this area did not have the problem of water pollution. It is with the advent of industries in this area that El Jadida is facing water pollution.

Groundwater pollution has resulted in the design of monitoring and decision-making systems. These systems are supported by in the information systems that can manage spatial dynamics of pollution processes. Monitoring of groundwater pollution is created to estimate short and long term changes, to carry out risk assessment analysis, to develop models for prediction and optimisation, and finally, to print maps for the support of decision-making processes. A number of automatic and manual-operated stations, chemical analyses in laboratories, satellite technology (Human use), risk assessment analyses and modelling systems is applied to improve our knowledge about behaviour of various types of pollution and its influence on water quality. Mapping of pollution together with their sources and the aquifer helps to illustrate all the phenomena more transparently.

New mapping approaches supported by a geographic information system (GIS) can be integrated into the standard user interface. Obviously, GIS has become an essential tool for providing boundary conditions to environmental models (Robin, 2002; Gourmelon and Robin, 2005). It becomes an indispensable instrument of essential importance in revealing associations between environmental exposures to hazardous substances and their impact on human health. The aim of a GIS is to provide the user with highly processed final products which are usually the information for *decisions* by which land administrative activities should be guided. The 'meaningful' information is considered as essential in a GIS and should be initially input from data sources.

GIS capabilities are mainly focused on spatial data management that includes a wide range of extensions for spatial analyses, development of three-dimensional map objects, network optimization, image processing, etc. New data inputs represent satellite and aerial images processed by remote sensing, and geopositional systems-GPS. Integration of digital tools for mapping of environmental pollution together with standard monitoring systems and environmental measurements illustrates figure 1. In this study GIS is used to correlate the data about heavy metal contamination, proximity to industrial fields and agricultural area.

The aim of this research is to examine groundwater pollution caused by industrial effluent and establish spatial relationship between groundwater constituents and pollution sources using GIS. A multivariate statistic approach was adopted to assist the interpretation of geochemical data.

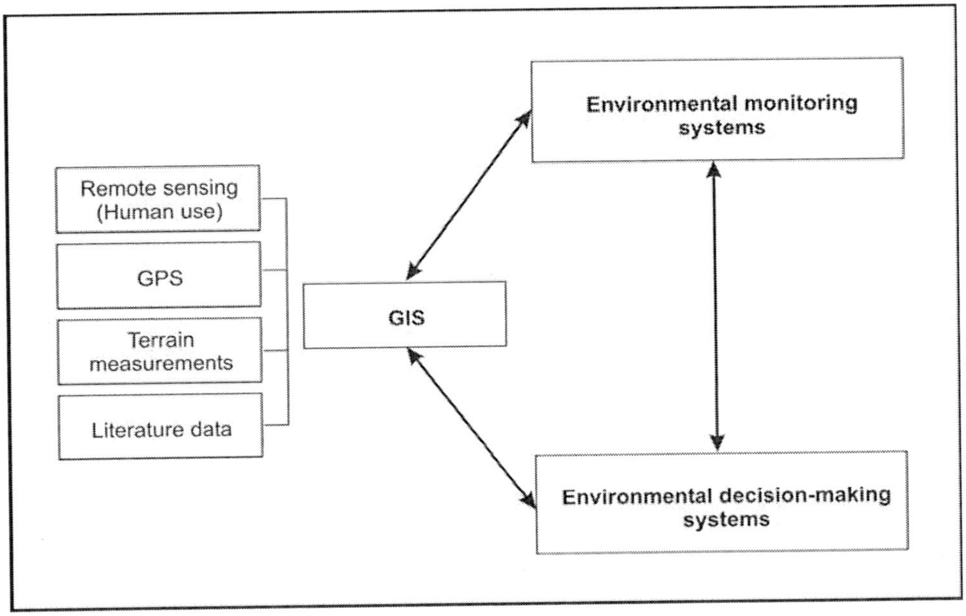

Figure 1. Integration of GIS tools in the frame of environmental policy.

Study Area

The study area, bounded by latitudes 33°00' and 33°12' N and longitudes 8°30' and 8°40' W, is located about 10 km south of El Jadida. The villages in the study area have a population of about 45,780 and a cattle population of about 19,200 (2004). Traditionally the people are an agricultural community. Groundwater in the basin is utilized for drinking, domestic and irrigational purposes. However there is a rapid industrial development in the basin along with increasing population initiating the necessity to evaluate groundwater resources (fig. 2).

The main industries phosphochemical, electricity, heavy engineering, chemical and ferrous metals factories established over the last two decades. Most of the industries are not having any sewer lines. Even today most of them don't have proper wastewater treatment plants. They generate approximately 4 million liters of effluent water per day, most of which is directly discharged to the natural coastal system, thereby causing enormous contamination of air, water and sediment, the degree of contamination has been so intense that in some parts the environment has become unsuitable for marine organisms living (Maanan, 2008).

The study area was blessed with abundant good quality groundwater during the pre-industrial period in Jorf Lasfar. Several studies have reported that the potential of groundwater contamination is increasing by excessive use of fertilizers and pesticides in the agriculture sector (El Achheb et al., 2002; Lhadi et al., 1996b), urban sewages (El Falaki and Lhadi, 2001) and leachate from municipal landfill sites (Chofqi et al., 2004). Groundwater involves intermingling between fresh water and ocean water, underground water quality is also becoming more saline due to excessive pumping (Lhadi et al., 1996a). Results of analyses indicated that there was no significant industrial pollution.

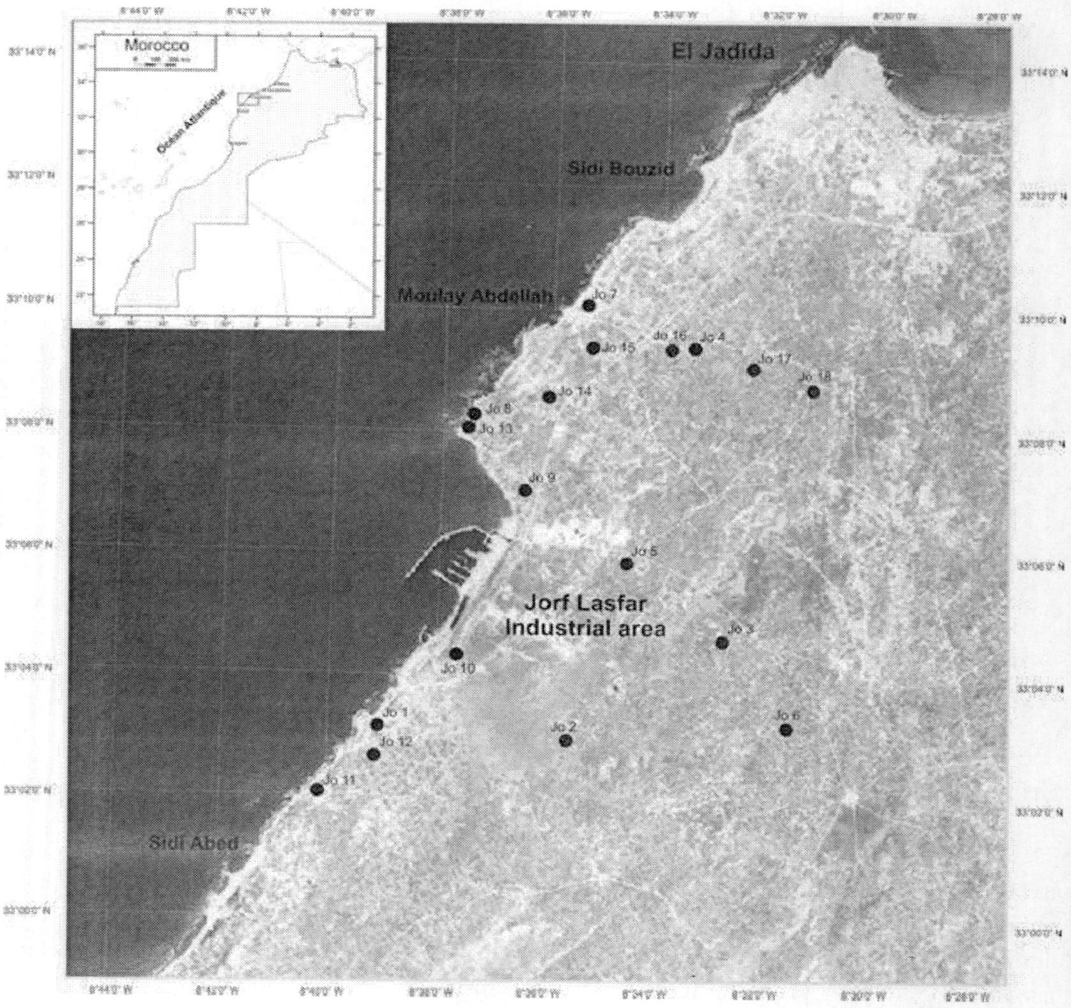

Figure 2. Wells sampling location.

The Jorf Lasfar discharges their waste effluents into the coastal area from various industries. Data on the following effluent characteristics of the major industrial sewage are given in Table 1: hydrogen-ion concentration (pH), total dissolved solids (suspended matter), chemical oxygen demand (COD), and chemical elements. Most of the effluents are hazardous in nature and are highly acidic to alkaline. The acidic pH of the effluents is mainly due to the presence of mineral acids, weakly dissociated acids, and salts of strong acids and weak bases. Electro-plating and lead-battery industries are some of the activities responsible for the acidic nature of the effluents.

Table 1. Principal industrial waste-effluent characteristics (2004)

pH	EC (µs/cm)	SM (mg/L)	COD (mg/L)	Concentration (mg/L)												
				Na^+	Cl^-	K^+	Ca^{2+}	Mg^{2+}	PO_4^{3-}	F^-	SO_4^{2-}	Cu	Cd	Fe	Zn	Pb
2,05	81200	824	813	1731	24630	359.1	1492	1508	201.5	183.5	9741	7	0.209	6.987	5.706	0.397

The climate is temperate. Precipitation is the main source of groundwater recharge mostly during November to February. The annual average rainfall, estimated from 1977 to 1998, is about 390mm, with a maximum in December and no rain during the dry period. The annual estimate of evaporation minus precipitation is 650 mm. The predominant wind directions are WSW to NW during the wet season and NNE to NE during the dry season. Sporadic violent winds (*Chergui*) occasionally blow from the ENE during the dry period and may contribute to high evaporation rates over the lagoon, as well as to extreme air temperatures, which can reach 40°C.Winds are predominately easterly and north-easterly parallel to the coast.

Geology and Hydrogeology

The regional geology of the El Jadida province comprises a folded Pre Cambrian/Lower Palaeozoic basement overlain successively by Mesozoic, Tertiary and Quaternary strata. The basement which outcrops in the south of the area includes schist, dolomites, siltstones, granites and rhyolitic lavas. The succeeding Triassic basalt lavas and Jurassic sandstones outcrop mainly along the banks of the Oum Rbia River (40 km north). The region is characterised generally by a series of broad plateaux varying in elevation from 125m to 200 m above mean sea level. Jorf Lasfar area is located along the western coastal margin of the El Jadida Plateau to the south-west of El Jadida. The western edge of the plateau exhibits a typical karstic terrain at an elevation of 50 to 60 m some 2 to 5 km inland. This terrain comprises mainly bedded limestone with only a thin soil cover due to extensive wind erosion. To the north-west the ground level falls gradually towards the Atlantic Ocean. A relatively steep slope is developed about 0.5 km from the shoreline and represents a former cliff line of Plio-Quatemary age.

The Cretaceous strata are subdivided into a lower division of Neocomian age and an upper division of Cenomanian age. The Neocomian is represented by red calcareous clays which occur north east of Safi and west of Mechra-Benabbou. The Cenomanian (Upper Cretaceous) strata are confined mainly to the El Jadida Plateau. The strata comprise a local basal conglomerate and an overlying sequence represented by calcarenites, marly limestones, marls and quartz sandstones. The Tertiary strata occur as a broad north-east to south-west trending strip up to 20 km wide which broadens in the vicinity of Safi and comprises mainly sandy marls and red clays. The succeeding Quatemary deposits predominate to the east of the Tertiary outcrop and also occur as a thin strip along the coast. The deposits include uncemented and cemented calcareous dune sands and calcretes, together with alluvial muds along the permanent river courses. The karstic terrain is also characterised by a thin surficial layer of tufa which has been precipitated on much of the limestone outcrop (fig. 3).

The Jorf Lasfar area is excavated within the Cenomanian limestones and marls of the El Jadida Plateau. These limestones are the major aquifer of the region. Available literature suggests the presence of significant groundwater, as confirmed by the presence of local groundwater wells and spring issues, notably "Source de Cap Blanc", which recorded a flow rate of 3 l s^{-1} (Ferré and Ruhard, 1975). Tests at Sidi Bouzid near El Jadida derived permeability of $5x10^{-6}$ to $5x10^{-4}$ m s^{-1}. Salinity levels generally increase towards the coast, as expected. However, there are anomalies within this trend, which may be a function of well depth or sample depth (saline water is more dense and salinity therefore increases with depth in the aquifer) or the presence of a direct conduit, eg. enlarged fissures typical of karstic terrains, between the well and the coast.

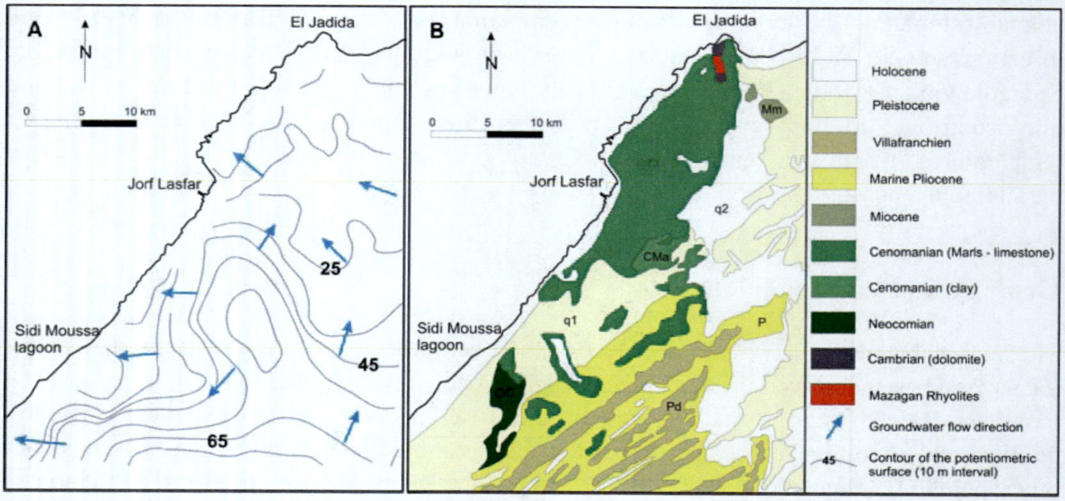

Figure 3. Hydrogeology and geology characteristics.

Materiel and Methods

Sampling and Analysis

18 georeferenced groundwater samples from open wells were collected in duplicate. Most samples were collected at locations where the local people complained of the brackish taste of water due to proximity of the area with sea (fig. 2).

A multiparametric probe designed for oceanic measurements in seawater is used. The probe was adapted to measure in situ physico-chemical parameters as temperature (T, °C), electric conductivity (EC, mS/cm) and pH while all other parameters were determined in the laboratory.

In the laboratory, the Ca^{2+}, Mg^{2+}, HCO^{-}_3 and Cl^{-} were determined by titrimetric method of analysis, while Na^+ and K^+ were analyzed by flame photometric method. The SO_4, P_2O_5 and NO^{-}_3 were analyzed by spectrophotometric method and a saturated calomel electrode (ISE25F) from Radiometer-Analytical S.A was used for determination of F concentration.

The water samples were analyzed for Al, Fe, Mn, Cu, Pb, Ni, Cr and Cd using Flame Atomic Absorption Spectrometry (A.A.S.) (ATI-UNICAM 929, Unicam Absorption Atomic, Cambridge CB12SU, UK) according to standard methods (Association Française de Normalisation (AFNOR, 1983). The EC was measured in microsiemens per centimeter at

25°C (μs/cm at 25°C). Except pH and EC, all other parameters are expressed in milligram per liters (mg/L).

GIS Approaches

The methods implanted in the GIS for the support of mapping of environmental pollution are mostly focused on data collection and basic analyses, modelling and predictions. The results can be displayed by the GIS in the map themes. The power of a geographic information system (GIS) within the framework of spatio-temporal analysis depends on its ability to manage a wide range of data formats, which are represented by digital map layers extended by attributes with various observations, measurements and pre-processed data. In addition to the standard functions (coordinate transformation, conversion between spatial formats, raster algebra, vector operations in the frame of geo-processing, network analysis, display and visualization), GIS contains development tools and software libraries that can be used to create program modules for solving basic data analysis and dynamic models.

ArcGIS software (ESRI, 2007) was utilized for the GIS analysis in this study. ArcGIS is a collection of software modules that create, edit, import, map, query, analyze, and publish geographic information. Geochemical data were mapped via satellite-generated coordinates (GPS bearings) for ease of future reference. A database was constructed matching GPS data to water data on heavy metals gathered at 18 sampling wells in and around Jorf Lasfar industrial area.

Geostatistical Analyst uses sample points taken at different locations in a landscape and creates (interpolates) a continuous surface. It provides two different groups of interpolation techniques: deterministic and geostatistical (Robin, 2002). All methods rely on the similarity of nearby sample points to create the surface (Robin, 2002). However, while deterministic techniques only use mathematical functions for interpolation, geostatistics relies on both statistical and mathematical methods (Cressie, 1993; Goovaerts 1997). Geostatistical techniques have been selected in this work, because they produce not only prediction surfaces but also error or uncertainty surfaces, giving an indication of how good the predictions are. The prediction maps were generated using ''ordinary kriging.'' This process involves a set of statistical methods that sample the nearest neighbors of a data set and use the results to interpolate a surface area that represents a prediction of the values based on a theoretical data point that lies between a set of actually measured data points.

Multivariate Analysis

Raw data were stored in MS Excel, and basic statistical parameters were calculated to acquire the overall feature of the data sets. The tests for normality of the raw and transformed data were performed using Statistica software, prior to multivariate analyses. To classify elements into groups that share similar geochemical features, multivariate analyses of cluster analysis (CA) and principal component analysis (PCA) were carried out using the statistical software package Statistica, Release 5.1 (StatSoft, 1997).

Results

1. Physico-Chemical Parameters

The chemical quality of groundwater is influenced to a considerable extent by the chemical composition of rocks and soil masses through which it moves under various physico-chemical conditions (Hansen et al., 2001; Grassi and Cortecci, 2005; Mulligan and Charette, 2006; Oberdorfer et al, 2008).

The pH values in the area range from 7.4 (table 3, well no. Jo8, Jo10 and Jo15) to 8.1 (table 3, well no. Jo4) in the shallow aquifer. The mean value of pH in the area is 7.6. Water are found to be moderately alkaline and typical of carbonate aquifers but within the USEPA (table 2) permissible limits (6.5-8.5). This alkalinity is advantageous since it would minimise the mobility of any heavy metals which may potentially be present in future ash placement. The electrical-conductivity values range from 85 μS/cm (table 3, well no. Jo7) to 948 μS/cm (table 3, well no. Jo13). The mean value is 347 μS/cm. The large variation is attributed to anthropogenic activities and to geochemical processes prevailing in the surface and subsurface regions.

Table 2. Maximum contaminant level (MCL) and secondary maximum contaminant level (SMCL) prescribed by the USEPA and WHO Guideline of different elements for drinking water.

Element	MCL	WHO
Cd	0.005	0.005
Cr	0.1	
Pb	0.015	0.05
Ni	0.1	
Element	**SMCL**	
Cu	1	1
Fe	0.3	0.3
Mn	0.05	0.1
Al	0.05 to 0.2	0.2
F	2	1.5

2. Cation Chemistry

Among the cations, Na^+ is the most dominant, constituting about 18% of the total. High concentrations of Na^+ must have resulted from the intense cation exchange reaction with Ca^{2+} and carbonate precipitation under alkaline conditions. High concentrations of Na^+ and Ca^{2+} in some sample are attributed to cation exchange among minerals and to the urban sewage contamination of Moulay Abdellah village. In the study area, the Na^+ concentration in groundwater ranges from 91.8-1456.2 mg/L. Groundwater in the carbonates terrain derive calcium from limestones rock. Calcium is next in abundance to sodium. As a result of the extremely mobile nature of calcium in the hydrosphere, it forms one of the major constituents of groundwater.

The source of magnesium (Mg^{2+}) is due to ion exchange of minerals in rocks and soils by water. The upper concentration limit for domestic purposes is about 268.8 mg/L. The concentration of Mg^{2+} in groundwater ranges from 21.2–232.7 mg/L. The presence of gypsum is major source of potassium (K^{+}). Very high concentrations of potassium (table 2, wells 1, 2, and 5) are influenced by the inflow of the industrial effluents. However, some wells are located in the agricultural area and the high content of potassium is attributed to the increased usage of potassium fertilizers. Cations such as magnesium and calcium and corresponding anions bicarbonate and sulphate are commonly added to groundwater by solution from carbonate aquifers (Todd, 1980).

3. Anion Chemistry

The anions constitute more than 35% of the total dissolved solids. Chloride is the dominant anion, followed by bicarbonate (18%) and sulphate (12%). The main source of chloride in groundwater in the area is the result of erosion and the sea containing brackish water (Andersen et al., 2005; Capaccioni et al., 2005). However, abnormal concentrations of chloride result from contamination by industrial wastes and from saline residues in the soil. Chloride ranges from 226.5-2842.8 mg/L in the coastal area. The source of most of the bicarbonate in the water is from carbonate aquifers and various human activities. The bicarbonate (HCO_3^-) of the region is highest in surface-water, with an average of 341.6 mg/L.

In pure water, nitrate (NO_3^-) is seldom present. In this area, nitrate concentrations range from 21.8–228.8 mg/L. The discharge of fresh groundwater from a coastal aquifer into the adjacent coastal marine environment varies strongly at a detailed scale along the coastline and may have impact on marine ecology. Nitrate pollution of groundwater may stem from different sources, including fertilizers, animal waste and domestic and industrial discharge (Antonakas and Lambrakis, 2000).

Sulfate derived from air pollutants widely contaminates the groundwater, and fertilizer and household wastewater also contribute high concentrations of SO_4^{2-}. The high concentrations at some places are attributed to the contamination by untreated industrial waste effluents. The fluoride content in groundwater ranges from 0.46–2.25 mg/L. The high fluoride content in groundwater results from the industrial pollution. In some of the villages "in south eastern of industrial area" where the water has high fluoride content, the people are afflicted with dental and skeletal deformities.

Fluoride distribution show high concentration in south-eastern of industrial area (1.91 mg/l) than Maximum allowable concentration (1.5 mg/l). The high groundwater fluoride values seem to be associated with the chemical industry plants that produce phosphate derivates including phosphoric acid and fertilizers, together with by-products such as sulphuric acid. The liquid effluents of the plants are discharged directly into the sea (183.5 mg/l) moreover, an alkaline pH is favorable for F^- dissolution (Sexena and Ahmed, 2003). In the present case, all shallow groundwater samples were weakly alkaline, with pH 7.6–7.8.

Table 3. T°, pH, EC and element concentrations in groundwater (all concentrations are in mg/l)

Puits	T° (°C)	pH	EC (ms/cm)	HCO_3^-	K^+	Mg^{2+}	Na^+	Cl^-	NO_3^-	$P_2O_5^+$	SO_4^{2-}	Ca^{2+}	Al	Fe	Mn	F	Cu	Pb	Ni	Cr	Cd
Jo 1	21.2	7.8	2.1	280.6	7.8	73.2	320.8	594.6	41.8	7.22	253.35	114.00	0.2820	0.0108	0.0014	1.8750	0.0049	0.0490	0.0250	0.0690	0.0075
Jo 2	19.9	7.7	2.9	366.0	11.8	191.5	1222.3	2387.4	187.2	6.99	447.88	272.80	0.2098	0.0188	0.0055	1.7580	0.0077	0.0513	0.0630	0.0630	0.0029
Jo 3	20.4	7.6	2.9	402.6	2.8	74.4	463.8	905.3	41.8	8.80	246.56	158.00	0.0695	0.0096	0.0096	1.9500	0.0125	0.0345	0.0583	0.0540	0.0025
Jo 4	20.6	8.1	2.4	213.5	2.3	51.6	181.9	461.5	146.9	5.08	183.22	90.00	0.0123	0.0160	0.0035	0.4100	0.0026	0.0290	0.0041	0.0065	0.0064
Jo 5	21.3	7.7	6.1	317.2	4.7	84.7	377.3	686.9	49.7	3.40	185.48	78.80	0.1460	0.3004	0.0177	1.5420	0.0096	0.0250	0.0610	0.0430	0.0045
Jo 6	21.4	7.6	0.9	305.0	6.5	14.4	337.4	665.6	63.4	7.50	882.18	340.00	0.0437	0.0194	0.0014	0.5500	0.0121	0.0392	0.0043	0.0073	0.0059
Jo 7	20.9	7.5	0.9	384.3	50.7	72.0	265.5	603.5	211.7	8.10	161.73	106.00	0.0560	0.0210	0.0032	0.6700	0.0153	0.0310	0.0461	0.0678	0.0061
Jo 8	21.0	7.4	4.0	427.0	10.9	100.8	722.5	1349.0	80.6	8.70	411.69	192.00	0.0497	0.1129	0.1613	1.3730	0.0016	0.0280	0.0740	0.0795	0.0049
Jo 9	20.1	7.5	4.0	402.6	8.6	90.0	357.9	647.9	136.1	9.60	122.15	102.00	0.0113	0.2067	0.1479	1.5870	0.0029	0.0270	0.0570	0.0647	0.0052
Jo 10	22.0	7.4	4.2	341.6	6.7	43.2	113.0	266.3	1.4	15.00	27.80	80.00	0.0101	0.1836	0.1297	1.5960	0.0020	0.0430	0.0482	0.0624	0.0075
Jo 11	21.3	7.7	3.3	286.7	9.8	69.6	445.6	816.5	62.6	7.24	275.97	134.00	0.2620	0.0104	0.0020	1.8200	0.0055	0.0390	0.0451	0.0672	0.0073
Jo 12	20.6	7.9	1.4	262.3	3.1	33.4	103.5	221.9	127.4	7.19	177.57	126.40	0.2780	0.0106	0.0021	1.8510	0.0057	0.0330	0.0428	0.0631	0.0068
Jo 13	20.9	7.6	9.5	335.5	51.2	196.8	1862.2	2813.4	80.3	8.50	567.76	170.00	0.0513	0.1210	0.1630	0.6970	0.0024	0.0361	0.0751	0.0802	0.0052
Jo 14	20.6	7.7	1.7	366.0	12.0	45.8	171.2	328.4	74.2	7.29	99.53	96.00	0.0352	0.0499	0.0560	0.5977	0.0065	0.0230	0.0050	0.0313	0.0043
Jo 15	20.9	7.4	4.4	378.2	4.7	107.0	661.8	1091.6	162.0	7.09	212.63	101.60	0.0373	0.0193	0.0033	0.5833	0.0111	0.0303	0.0054	0.0074	0.0062
Jo 16	20.2	7.8	1.3	268.4	7.8	70.8	85.7	257.4	39.2	5.42	156.08	86.00	0.0120	0.0260	0.0054	0.5800	0.0106	0.0420	0.0052	0.0071	0.0065
Jo 17	20.8	7.6	2.1	366.0	47.7	75.6	176.7	381.6	66.2	4.78	28.72	52.00	0.0160	0.0420	0.0048	0.4800	0.0011	0.0140	0.0029	0.0053	0.0062
Jo 18	20.4	7.6	8.4	207.4	10.0	268.8	934.2	2449.5	68.8	6.79	628.84	274.00	0.0428	0.0258	0.0072	1.2150	0.0068	0.0150	0.0306	0.0028	0.0044
Min	19.9	7.4	0.85	207.40	2.3	14.4	85.7	221.9	1.4	3.40	27.80	52.00	0.0101	0.0096	0.0014	0.4100	0.0011	0.0140	0.0029	0.0028	0.0025
Max	22.0	8.1	9.48	427.00	51.2	268.8	1862.2	2813.4	211.7	15.00	882.18	340.00	0.2820	0.3004	0.1630	1.9500	0.0153	0.0513	0.0751	0.0802	0.0075
Mean	20.8	7.6	3.47	328.38	14.4	92.4	489.1	940.5	91.2	7.48	281.62	142.98	0.0903	0.0669	0.0403	1.1742	0.0067	0.0327	0.0363	0.0434	0.0056

One of the mechanisms of high F^- concentrations in groundwater in arid and semiarid regions is the condensation of soluble components due to evaporation and evapotranspiration (Jacks et al., 2005). As consequent, certain forms of human and animal morphological anomalies (dental and skeletal fluorosis) were observed in southern of industrial area (Sidi Abed) due to the excessive ingestion of fluoride in water drinking.

4. Heavy Metals Distributions in Groundwater Samples

The main objective of this study is to evaluate the impacts of industrial area of Jorf Lasfar on groundwater in terms of heavy metal contaminations. Groundwater samples in the natural slopes are assumed to be free from any anthropogenic contaminations and thus they could be used to evaluate the degree of heavy metal and trace element contaminations in the industrial area.

Groundwater heavy metals constituents were analyzed by considering the mainly USEPA (2003) standards for drinking water purposes (table 2). The permissible and desirable limits for presence of various elements in groundwater are given in the table 3.

The range of iron is found to vary between 0.01 to 0.3 mg/l. The Fe concentration of groundwater is determined as 0.3 mg/l in the wells no. Jo5. Fe concentration in this region is higher than USEPA limit value of 0.3 mg/l. The highest manganese concentrations are found as 0.2 mg/l. For the study area, it has been found that in 4 locations the manganese concentration exceeds the USEPA limit. The locations of Mn polluted areas show that the source of Mn pollution is the traffic and the unsanitary deposits in Jorf Lasfar area.

The Cd and Cr content in water are generally higher at well near industrial area, particularly those characterized by seawater intrusion. The strong correlation of Cd with Cr ($r = 0.83$) with a low correlation for Cu ($r = 0.64$) in the groundwater suggests the same origins and sources. The manufacture of phosphate fertilizer in Jorf Lasfar results in a redistribution of the Cd and Cr in the rock phosphate between the phosphoric acid product and the gypsum waste.

Nickel concentration of groundwater within the study area does not exceed the limit value of 0.1 mg/l but in some locations Ni concentration is quite high and it is found to be 0.06 mg/l. The high Ni values are found near Jorf Lasfar harbour.

Lead ranges between 0.0140 and 0.0513 mg/l, the high value of lead were detected in south of industrial area “along the principal wind direction”. Aeolian influx of anthropogenic Pb originating from metallurgical processing in the Jorf Lasfar has also been detected as responsible of Pb pollution in the groundwater. The release leaded gasoline from the 1950s to the 1970s in North America and until the 1980s in Europe (Candelone and Hong, 1995) reached such a magnitude that it became readily detectable by increasing total Pb contents in various kinds of natural monitors (Nriagu and Pacyna, 1988).

5. Multivariate Analysis

Cluster Analysis was used to classify elements into groups with good correlations within the groups (Zhang, 2006). The Pearson correlation coefficients were selected as the measurement between groups, and the cluster method in use was furthest neighbour, in order to make sure that variables classified into the same group shared good correlations. The dendogram in Fig. 4 shows the results of applying a nearest neighbour aggregation strategy. In this strategy, the two elements presenting the nearest distance in the matrix will form the first group or class. The distance from a new element to a group already formed will be taken as the nearest distance between that element and its closest element in the group.

The first level of aggregation is established with Cu-Cd, Ni-Cr and Pb which, at the same time, forms a cluster with the Mn, Fe, Al and EC. These elements are associated with each other in the next stage and later with F^-. Finally, $P_2O_5^-$ is added, forming a cluster with the previous ones on the right side of the dendogram.

A second cluster, on the left side of the dendogram, shows a first level of association with the pair Mg^{2+}–NO_3^-, as the nearest variables. Then, Ca2+, very close to the previous elements, is added, and later on, $SO^{2-}{}_4$, $HCO^-{}_3$, Na and Cl.

The dendogram in Fig. 4 shows the association of all analyzed variables in two big families. They represent, in each case, the variables that quantify the concentration of typically marine indicators (Na, Cl and Ca) on the one hand, and typically industrial ones (Mn, F, Pb, Ni, Cr, Cd and Cu) on the other.

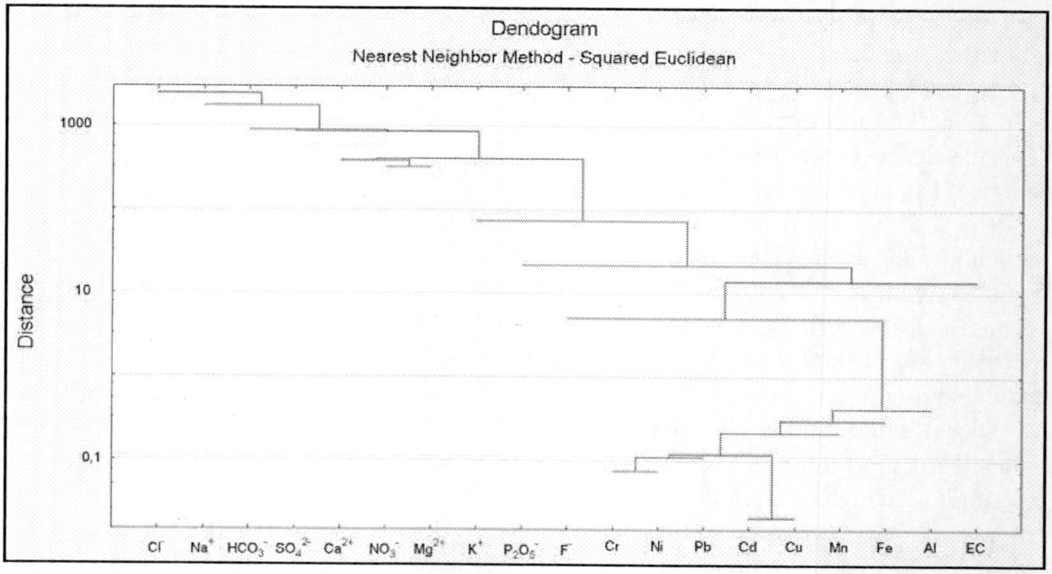

Figure 4. Cluster tree of elements using cluster analysis based on Pearson's correlation coefficients.

6. GIS Analysis

The GIS mapping technique was employed to produce the spatial distribution maps for the eight observed pollutants in groundwater. To obtain the overall patterns of these elements, the spatial interpolation method of inverse distance weighted (IDW) was applied with neighbouring samples used for estimation of each grid point. The power of one was chosen to acquire some degree of smoothing effect. Good spatial patterns for all these elements were thus obtained. Other more complicated spatial interpolation methods such as kriging may be considered. However, in this study, the objective is to identify the overall pollution pattern, and the relatively easy method of IDW is sufficient for such a purpose. It should be mentioned that there are uncertainties in the results of any spatial interpolation method, and results on specific locations should be regarded as the expected possible values, which are not necessarily the true values that in theory can never be obtained.

The heavy metal contents data were used for the spatial analysis and integration for groundwater quality mapping. Overlay analysis was performed for generating a water quality map where the various quality maps were overlaying these maps the polluted areas such as seawater intrusion area, agriculturally and industrially areas are determined.

The geochemical maps of Al, Fe, F, Mn, Cu, Ni, Cr and Cd are presented in fig. 5 and fig. 6. In these maps, the concentration ranges were decided with reference to USEPA limit values of different elements for drinking water (USEPA, 2003).

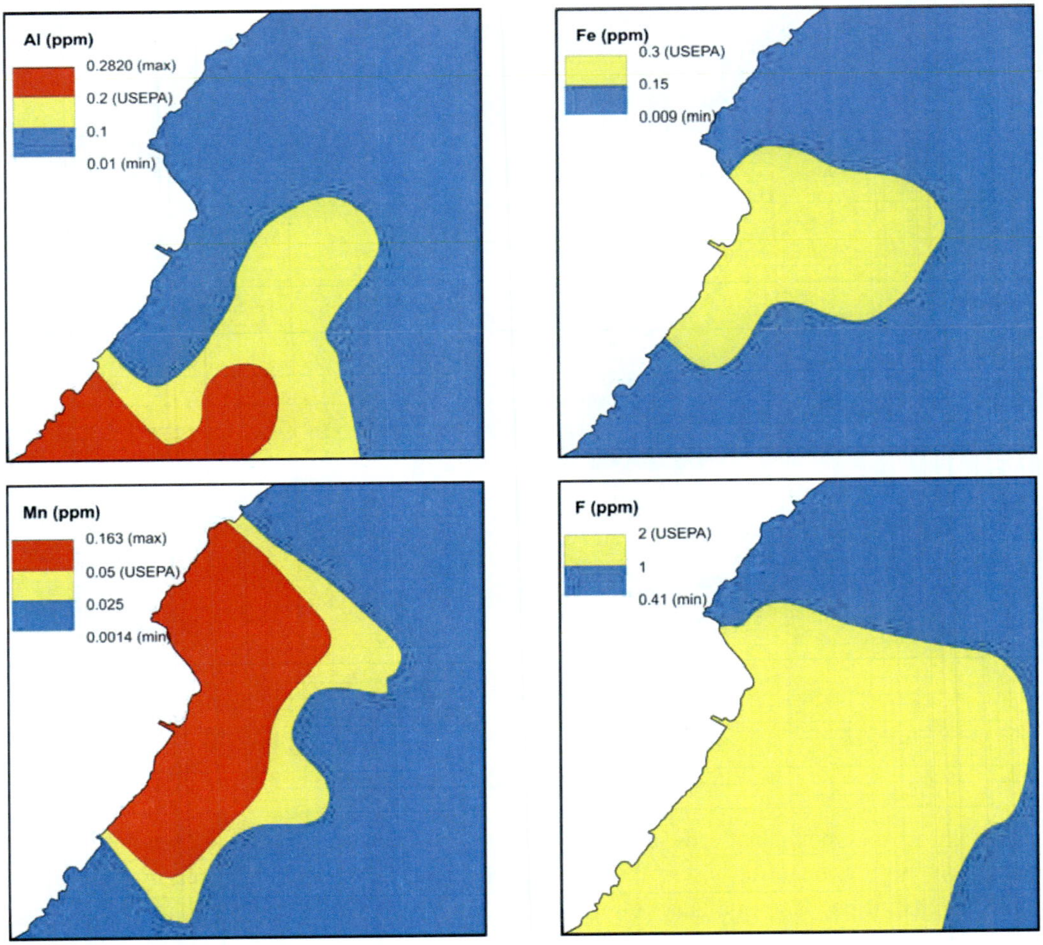

Figure 5. Al, Fe, Mn and F distribution.

In general, several hot-spots of high metal concentrations were identified in the geochemical maps in particular in south-west of industrial area. Fig indicates that the spatial distribution of the impact degrees of each pollution source were significantly different from each other. Each of the eight elements Al, Fe, F, Mn, Cu, Ni, Cr and Cd show elevated concentrations in groundwater in the south of industrial area. Al shows high concentrations extending towards the west side and coastal area of Sidi Abed village. A similar picture is shown for Mn and Cd. Mn has also shown high values along the coastal area, and Al and Cd also exhibit such a feature for some locations. The element Cd illustrates relatively high values along the coast and in the Jorf Lasfar industrial area.

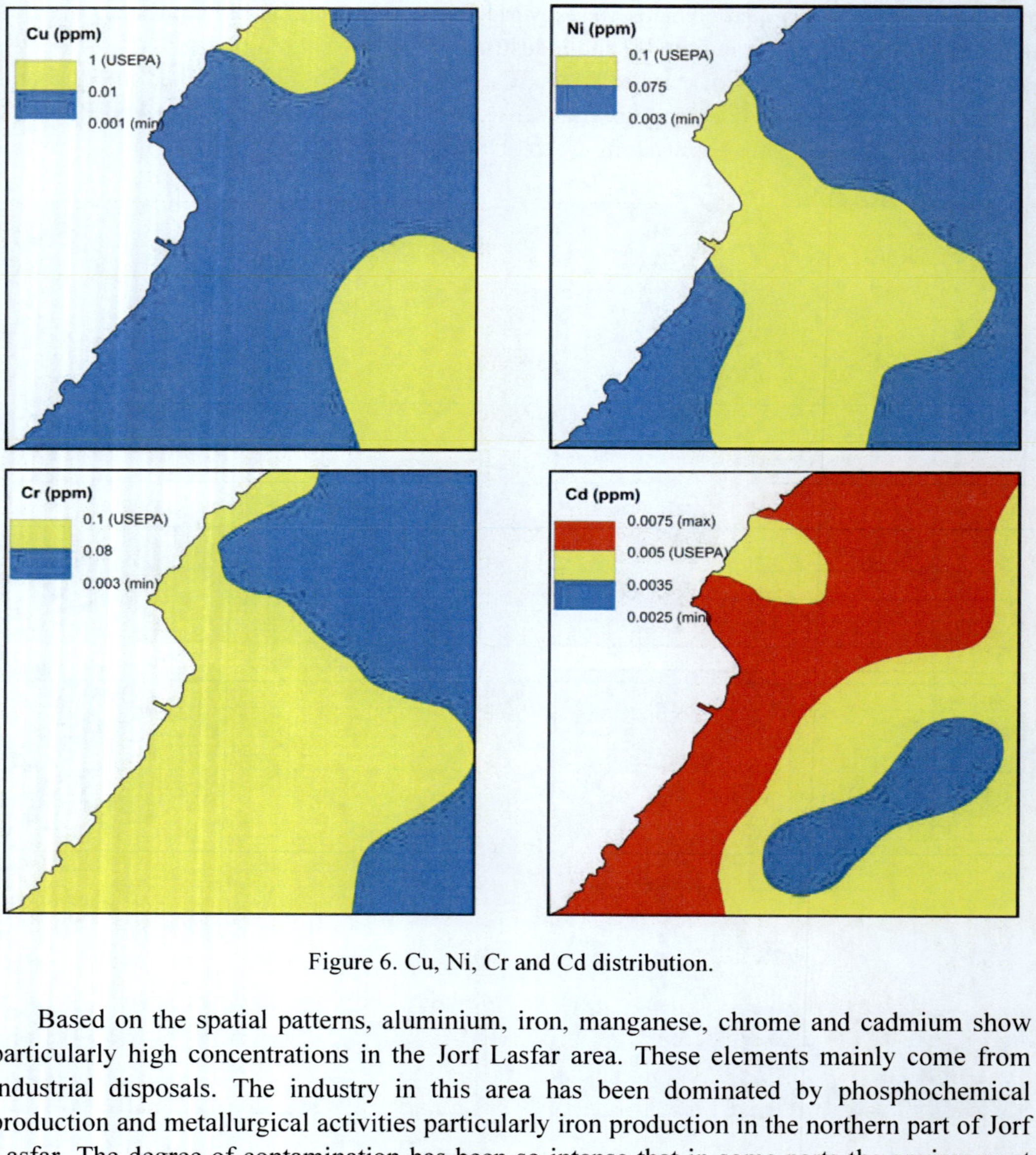

Figure 6. Cu, Ni, Cr and Cd distribution.

Based on the spatial patterns, aluminium, iron, manganese, chrome and cadmium show particularly high concentrations in the Jorf Lasfar area. These elements mainly come from industrial disposals. The industry in this area has been dominated by phosphochemical production and metallurgical activities particularly iron production in the northern part of Jorf Lasfar. The degree of contamination has been so intense that in some parts the environment has become unsuitable for human living. The isolated pollution hotspots may imply local pollution sources that need to be further investigated.

Further studies may be carried out to investigate and confirm the hotspots, to identify other possible pollution sources, and to establish the relationships between pollution and human health of the residents. These studies become important with the rapid population growth and urban and industrial development in Jorf Lasfar area that has the potential to give rise to pollution of groundwater.

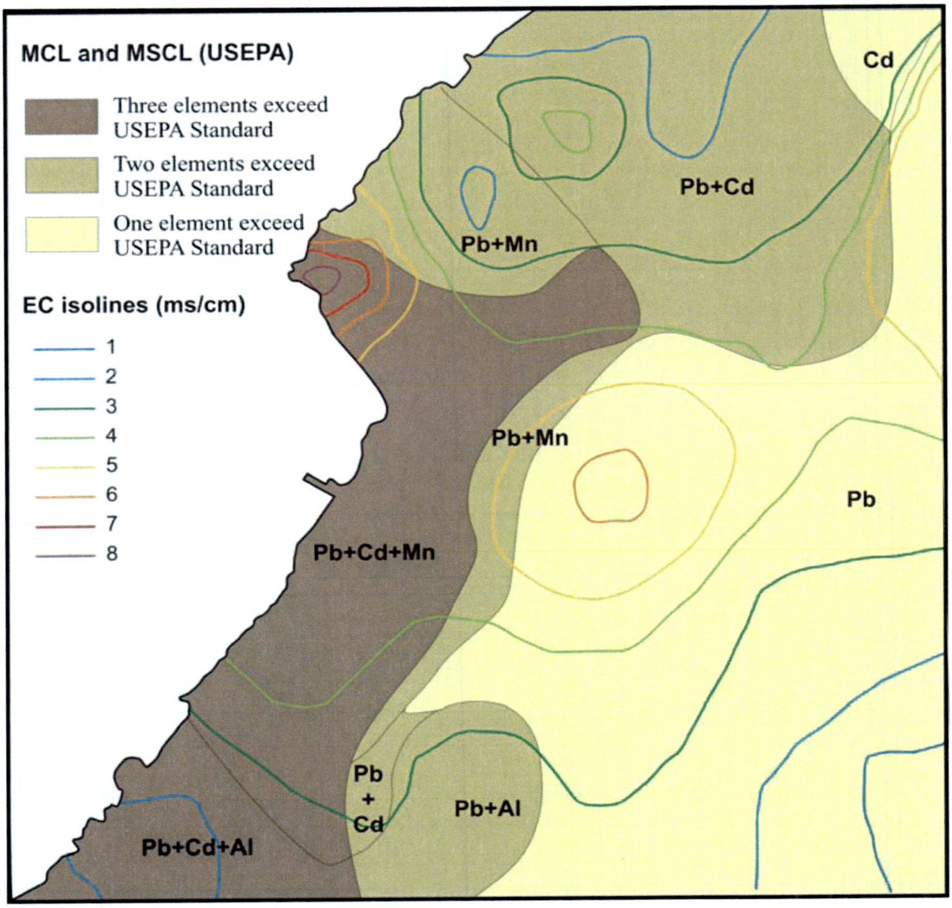

Figure 7. Mapping risk for drinking water.

Based on Maximum contaminant level (MCL) and secondary maximum contaminant level (SMCL) of different elements for drinking water prescribed by the USEPA and EC levels GIS is used to identify area with risk. The results of the sampling campaign presented in this paper allow confirming that the groundwater in this area is classified as pollute to high polluted located near industrial area. The majority of problems were identified near industrial area, particularly those characterized by seawater intrusion. The main problems related to drinking water quality are associated with the conditions of the water supply networks, the pollution of 'parent' water and in particular the contamination of groundwater with pollutants of both anthropogenic and natural origin, as well as the intrusion of seawater in aquifers.

Pollutants in groundwater aquifer may constitute a significant human health risk. A large variation in response may result among human populations experiencing the same level and duration of exposure to pollutants.

Conclusion

This paper presents the heavy metals concentrations in groundwater samples in the highly industrial area form Jorf Lasfar, Morocco. The results from the natural slopes were used to evaluate the impacts of industrialization on the heavy metal contents of groundwater in the developed areas.

The significant degradation in the quality of the groundwater in the area is indicated by the high concentrations of Na^+, Ca^{2+}, Cl^-, and HCO_3^- ions in samples from dug wells and bore wells located in the vicinity of the effluent-disposal sites. These wells penetrate fractures that have a relatively high hydraulic conductivity and rapidly transport contaminated water through the groundwater system by advective flow.

Multivariate statistic approaches were adopted for data treatment, allowing the identification of three main controlling the heavy metal variability on regional scale. Geostatistics were used to construct regional distribution maps, to be compared with the geographical, geologic, hydrologic, and land use regional database using GIS soft-ware. This approach, evidencing spatial relationships, proved very useful to the confirmation and refinement of geochemical interpretations of the statistical output. In general, the results confirm the contribution of both natural and non-point pollution to the chemical properties of groundwater.

Using ArcGIS software, spatial analysis and integration were carried out for mapping drinking water quality in the basin. From the distribution of heavy metal, it is inferred that the excess concentration of F, Fe, Ni, Mn, Mo and Cu at some locations is the cause of undesirable quality for drinking purposes. The reason for excess concentration of various heavy metals is the industrial activities and salinity levels show the sea water intrusion.

The results of the analysis indicate that the groundwater can not be used as drinking water. The electrical conductivity is high in some location so groundwater can also not be used as irrigation water. This study has demonstrated the importance of anthropogenic influences on the large-scale regional water quality around the Jorf Lasfar area. These anthropogenic influences are dominated by effluent inputs from industrial manufactories and from urban centres (Moulay Abdellah and Sidi Abed villages).

References

Andersen, M.S.; Nyvang, V.; Jakobsen, R.; Postma D. 2005. Geochemical processes and solute transport at the seawater/freshwater interface of a sandy aquifer. *Geochimica and Cosmochimica Acta* **69**, 3979-3994.

Antonakas, A. and Lambrakis, N., 2000. Hydrodynamic characteristics and nitrate propagation in Sparta aquifer. *Water Research*. **34**, 3977-3986.

Candelone, J.-P. and Hong, S., 1995. Post-Industrial Revolution changes in large-scale atmospheric pollution of the northern hemisphere by heavy metals as documented in central Greenland snow and ice. *J. Geophys. Res.* **100** (D8), 16.605-16.616.

Capaccioni, B.; Didero, M.; Paletta, C.; Didero, L. 2005. Saline intrusion and refreshening in a multilayer coastal aquifer in the Catania Plain (Sicily, Southern Italy): dynamics of degradation processes according to the hydrochemical characteristics of groundwaters. *Journal of Hydrology* **307**, 1-16.

Cressie, N., 1993. *Statistics for Spatial Data,* John Wiley & Sons Inc. 928 p.

El Achheb, A.; Mania, J. ; Mudry, J. ; Chauve, P. ; Ouaaka, A. 2002. Essai de bilan des flux azotés dans les eaux souterraines sous climat semi-aride (Cas des périmètres irrigués des Doukkala - Maroc). *Revue Française de Géotechnique* **101**, 105-111.

El Falaki, K. and Lhadi, E. K., 2001. Contribution à l'étude d'impact de l'épandage d'eaux usées brutes sur l'environnement dans la région de Zemamra (province d'El Jadida, Maroc). *Déchets sciences & techniques* **24**, 28-32.

ESRI, 2007. ArcGIS, version 9.2 Arc-Doc: Environmental Systems Research Institute, Inc., Redlands, CA., documentation at <http://webhelp.esri.com/arcgisdesktop/9.1>.

Ferre, and Ruhard, JP 1975. Les Bassins des Abda-Doukkala et du Sahel de Azemmour a Safi. *Notes Serv. Geol. Maroc* **231**, 261-298.

Goovaerts, P., 1997. *Geostatistics for Natural Resources Evaluation,* Oxford University Press. 482p.

Gourmelon, F. and Robin, M., 2005, *SIG et littoral*, Hermes Press. 332 p.

Grassi, S. and Cortecci, G., 2005. Hydrogeology and geochemistry of the multilayered confined aquifer of the Pisa plain (Tuscany – central Italy). *Applied Geochemistry* **20**, 41-54.

Hansen, L.K.; Jakobsen R. and Postma D., 2001. Methanogenesis in a shallow sandy aquifer, Rømø, Denmark. *Geochimica et Cosmochimica Acta* **65**, 2925-2935.

Jacks, G.; Bhattacharya, P.; Chaudhary, V. and Singh, K.P., 2005. Controls on the genesis of some high fluoride groundwaters in India. *Applied Geochemistry* **20**, 221–228.

Lhadi, E.K. ; Mountadar, M. ; Younsi, A. ; Martin, G.; Morvan J. 1996b. Pollution par les nitrates des eaux souteraines de la zone littorale de la province d'El Jadida (Maroc). *Hydrogéologie* **3**, 21-33.

Lhadi, E.K.; Mountadar, M.; Younsi, A.; Martin, G.; Morvan, J. 1996a. Contamination par les sels du systéme aquifére côtier de la province d'El Jadida (Maroc). *Hydrogéologie* **3,** 35-49.

Maanan, M., 2008. Trace metal contamination of marine organisms from the Moroccan North Atlantic coastal environments. *Environmental Pollution* **153/1,** 176-183.

Mulligan, A.E. and Charette, M. A. 2006. Intercomparison of submarine groundwater discharge estimates from a sandy unconfined aquifer. *Journal of Hydrology* **327**, 411-425.

Nriagu, J.O. and Pacyna, J.M. 1988. Quantitative assessment of worldwide contamination of air, water and soils by trace metals. *Nature* **333**, 134–139.

Oberdorfer, J.A.; Charette, M.; Allen, M.; Martin, J.B. and Cable, J.E. 2008. Hydrogeology and geochemistry of near-shore submarine groundwater discharge at Flamengo Bay, Ubatuba, Brazil. Estuarine, *Coastal and Shelf Science* **76**, 457-465.

Robin, M., 2002. La télédétection : Des satellites aux SIG. Nathan Press, 2nd edition 128 p.

Sexena, V.K. and Ahmed, S., 2003. Inferring the chemical parameters for the dissolution of fluoride in groundwater, *Environmental Geology* **43,** 731-736.

StatSoft, Inc. 1997. *Statistica for Windows*, Release 5.1. Tulsa, Oklahoma.

Todd, 1980. *Groundwater Hydrology*, second edition, David K Todd, Wiley & Sons, 535 pages, 1980.

USEPA, 2003. List of contaminants and their maximum contaminant level (MCLs). EPA 816-F-03-016, June 2003, http://www.epa.gov.

Zhang, C, 2006. Using multivariate analyses and GIS to identify pollutants and their spatial patterns in urban soils in Galway, Ireland. *Environmental Pollution* **142**, 501-511.

INDEX

A

B

C

D

G

H

I

N

O

P

Q

R

S

T

U

V

W

Y

Z